F. S Granger

Psychology

A Short Account of the Human Mind. Second Edition

F. S Granger

Psychology
A Short Account of the Human Mind. Second Edition

ISBN/EAN: 9783337365493

Printed in Europe, USA, Canada, Australia, Japan

Cover: Foto ©berggeist007 / pixelio.de

More available books at **www.hansebooks.com**

PSYCHOLOGY

A SHORT ACCOUNT OF THE HUMAN MIND

BY

F. S. GRANGER, M.A., D.Lit. (Lond.)

AUTHOR OF "THE WORSHIP OF THE ROMANS,"
PROFESSOR IN UNIVERSITY COLLEGE, NOTTINGHAM

SECOND EDITION, REVISED

METHUEN & CO.
36 ESSEX STREET, W.C.
LONDON
1897

Richard Clay & Sons, Limited,
London & Bungay.

PREFACE TO FIRST EDITION.

THIS book is intended as a brief introduction to the study of the human mind. In writing it, I have tried to anticipate and to remove some of the difficulties which are usually felt by students on taking up the subject for the first time. The illustrations are largely drawn from every-day experiences and familiar books. The use of technical terms has been avoided as far as possible. Where they occur, they are explained in the text, and referred to in the index by a different type. The index thus, to some extent, serves the purpose of a glossary. Those readers who do not know any logic, may find it advisable to begin the book at the second chapter, and to leave the first until they have read the account of logical processes given in the chapter on *Reasoning*.

I am indebted to Prof. Symes for some useful hints, and to Mr. J. W. Carr, M.A. for revising the proofs from page 65 to the end.

F. S. G.

University College, Nottingham,
    April 1891.

PREFACE TO SECOND EDITION.

In this edition I have made a few corrections, and have added one or two short paragraphs upon topics that seemed to need further explanation. In them I have dealt with the *organization of experience* (p. 55), *apperception* (p. 122), *psychical universes* (p. 147), and *the practical syllogism* (p. 230). These points were already dealt with implicitly. In the first edition of so elementary a text-book as this, it did not seem necessary to acknowledge the sources upon which I drew. I made one exception to this, and in the index against Wundt's name wrote *passim*. I have not, therefore, borrowed from him without acknowledgment. My debt is even greater to others who are incidentally named. In spite of some passing fashions, I am content to have drawn largely upon Aristotle, the *Ethics* of Spinoza, and Huxley's monograph upon Hume.

The additions that have been made, will not interfere with the use of this second edition, in class, along with the first.

F. S. G.

University College, Nottingham,
  August 1897.

CONTENTS.

CHAP.		PAGE
I.	ON PSYCHOLOGICAL METHOD	1
II.	GENERAL CONDITIONS OF MENTAL DEVELOPMENT	21
III.	SENSATION	56
IV.	LAWS OF MIND	79
V.	MEMORY	104
VI.	REASONING	123
VII.	PERCEPTION	148
VIII.	THE FEELINGS	179
IX.	WILL	198
	INDEX	231

PSYCHOLOGY.

CHAPTER I.

ON PSYCHOLOGICAL METHOD.

1. **Introduction.**—Psychology is the name given to the study of mind. Like other sciences, it starts from the familiar and the obvious, and is therefore foreshadowed in common ways of thinking; we are all psychologists, even if we do not know it. Those persons who are least given to reflection, habitually mark off the outer world of form and colour, of force and movement, from the inner world of thought, feeling, and will; the realm of nature from the life of the soul. In making this distinction, they anticipate the separation of the sciences concerned with mind from other sciences; while physics and chemistry, for instance, deal with changes in external things, psychology looks within, and observes the changes which take place there. Another familiar distinction is often made, namely, between being and seeming; and this, too, is interesting to the psychologist. Changes in outer objects, and indeed outer objects themselves, only become known to us through the impressions which they make upon us. The blue of the sky which shows through the rain-clouds, the sparrows

twittering in yonder trees, the cart which rumbles down the road, the clock ticking above my head, must affect me through one of the channels of perception, the eye or the ear for example, or else they might just as well not be at all, as far as I am concerned. Usually, indeed, we prefer to know what a thing is, and leave on one side as unimportant what it seems; the psychologist, however, is more concerned with the operation of things upon us, than with their actual nature; being yields in importance to seeming. The colours and sounds, the pressures and repulsions which surrounding things give rise to, and by which they make themselves known, are not viewed as the attributes of those things, but as the effects which they produce in us. We have to remember that to see or to hear anything is simply to have an impression of sight or hearing, and that these impressions only exist in the mind of the observer; and so that outer world, which generally seems so real, fades away, and leaves us in possession of a bundle of impressions. Side by side with these impressions of outer things, there pass across our inward view reflections, feelings, desires, which are present to us in the same immediate fashion as sense impressions. To the psychologist, these other elements of our mental states are also real in a sense which exceeds the reality of the outer world; they are known directly, the outer world is known only through its effects upon us. For the psychologist a purpose, or an emotion, is more real than the ground on which he stands.

2. **Use of terms in Psychology.**—One of the first questions in dealing with a science is this: What terms shall be used to describe its objects? The question is peculiarly important in psychology. As we saw in the last paragraph, the states with which it is concerned are the very material

out of which our knowledge is built up, and cannot therefore be defined by anything more familiar than themselves. It will be necessary, then, to start with the terms ordinarily used, and gradually to make clear what we mean by them.

Greater caution is required here than almost anywhere else in the use of terms; and not once, nor twice, has carelessness in this respect produced the most disastrous consequences. The use of names, like reason, conscience, will, for classes of mental operations, has given way to their use to denote the so-called faculties; and then, by a very comprehensible error, these faculty-names have been thought to explain the operations which they denote. Conscientious acts have been interpreted by a reference to conscience, volitions by a reference to the will, and so forth. Of course this does not help us; the mind has not its will here, its conscience there, and its reason somewhere else; it reasons, wills, and is conscientious as a whole; yet this fallacy is so ingrained in common speech that it is difficult entirely to free ourselves from it.[1]

The student of the standard psychological works of the present stands in little danger of such an obvious mistake as this; his perils are of a subtler kind. Among thinkers who have paid attention to psychology, we can generally trace the influence of their other studies on their style of thought and expression. Thus Kant and Locke still make use of the language of the schools with the corrections and explanations necessary for their purpose. The scientific bent of Descartes is shown in the deductive form of his *Principles of Philosophy*. Herbart, again, borrows from the rapidly advancing physics of his time the notion of a mechanism, and applies it to psychology; and perhaps the

[1] Spinoza, *Ethics*, ii. 48.

great importance attached by the Associationist school to the accumulation of small mental elements may be explained by the analogy of the Penny Savings Bank. Logic, mathematics, mechanics, and political economy having thus in turn contributed to the psychologist's vocabulary, it might have been expected that the brilliant discoveries which have been made of late years in biology would not fail to leave their mark, and this we find to be the case; expressions like function, development, and so on, are used in a biological sense.

It may not be unprofitable, then, to consider for a moment how far the use of such metaphors may affect our views of the facts. There is this risk at any rate, namely, that in employing any single set of terms for complex and varied groups of facts, some of them may be distorted or lost sight of. The ultimate feelings of our nature are probably of more than one kind, and, therefore, when we use the metaphors drawn from the sense of sight such as trace, image, background, or, from the sense of resistance, such as motive, conflict, tension, we must always remember that we are, after all, dealing with figures of speech, and that they have to stand for all kinds of experiences however received, and not merely for those received by way of sight or the muscular sense. More dangerous suppositions still lurk in the application of metaphors drawn from the commoner movements, as, for instance, grasping, weighing, apprehending, and so on; here we are led to think of the mind as a workman standing outside of, and having a separate existence from, his work. Physiological expressions are dangerous, too, when transferred from the nervous process to the mental one. Recently a physician, lecturing in one of the older universities to teachers, defined thought to consist in "the formation of the union of cells!"

It would almost seem, then, that the danger of confused habits of thought is less when the terms used in psychology are colourless and general in their associations. In this way we should be led to prefer the fairly abstract vocabulary of physics, or the quite abstract one of pure mathematics; and it is one of the chief services performed by Herbart that he clothed his system in terms of such general applicability. The reader will therefore be prepared to use algebraic symbols at times, as has so long been done in logic. We shall find as we proceed that the study of psychology involves the dissection or analysis of complex states of mind into their constituent parts; hence terms like element, factor, sum, combination, tendency, resultant, product, will be used with the signification which they possess in physics and pure mathematics. By the term element, particularly, will be meant any constituent part of a state of mind; by the term factor, will be meant any circumstance which contributes directly or indirectly to its occurrence. For example, some visual sensations are necessary constituent parts in a complete perception of a clock, and may thus be spoken of as elements in that perception. A necessary circumstance in my perception is that I should be in the neighbourhood of the clock. This, then, is a factor. Observe that all elements in a state of mind are necessarily circumstances which determine it, and are therefore factors. We may not turn the statement round and say that all factors in a state of mind are necessarily elements which we can observe in it. The signal point in the study of psychology is that the factors which determine any single state of mind are for the most part removed from our immediate inspection, and can only be arrived at by inference. Your esteem for one person, and aversion from another, are states of mind determined by many past experiences which are not clearly remembered, and do not therefore enter as elements into those states of mind, but for all that they have been factors in their production.

3. **Relation of Psychology to other Sciences.**—Psychology is a science with well-marked boundaries. Its subject-matter consists in those states which constitute our inner, or mental, life, as opposed to the changes in the outer world which are investigated by natural science. It uses the same instruments as they, observation and experiment, and the same methods, induction and deduction; but it has this advantage over them, that while they have to interpret sensations of resistance, colour, form, sound, and smell into those attributes of external objects, which give rise to them, as we saw at the beginning of this chapter, and so get their materials second-hand, the student of mind need not go further than his own consciousness for his subject-matter, and can question his own experience directly. It is in this sense that Aristotle, at the beginning of his work on this subject, speaks of its greater accuracy.

There are, of course, numerous points of connection between the study of mind and other sciences. In enumerating the conditions of mental action, account has to be taken of physiological conditions, and of all the facts in chemistry and physics which they involve. Further, the operations of the mind form one group of cases in which the laws of dynamics are illustrated, and thus depend for explanation on the sciences of force and quantity; and, to a certain extent, the different intensities and forms assumed by mental states, are capable of exact treatment by these sciences.

Psychology occupies a central position in the studies which are concerned with man. From it diverge, on the one hand, the sciences which regard him as one among the animals, and investigate his anatomical characteristics, his relation to other animals, the differences produced in his physical constitution by climatic and other conditions; on the other hand, the sciences which treat of him in his social relations, as a member of a body politic, as a wealth-producing and consuming being, and so on. In all these sciences he is regarded as conforming to some average type.

Special branches of science, however, are occupied with the divergencies from the average which arise from time to time. Pathology in its ordinary acceptation, deals with physiological deviations. Corresponding to, and in close connection with, this science, stands mental pathology which deals with aberrations like insanity, hypnotism, crime, and so forth. Psychology has lent its own light to solve the difficulties of these sciences of error, and has in turn received contributions from them. The study of insanity, hypnotism, dreaming, and crime, has already helped to solve many questions in psychology.

There are several subordinate sciences which stand in the closest relations to psychology. Each of the three chief aspects under which mental operations are most often regarded—thought, feeling, and will—has given rise to a special science. Logic investigates the forms of thought which subserve the attainment of the truth. Æsthetic considers those relations between the mind and its object which give rise to the feeling of the beautiful. Ethic scrutinizes that relation between the agent and his act on which moral approval is passed.

4. **The Instruments of Psychology: Introspection.**—Reference has already been made to the fact that psychology can deal with its material at first hand. Our states of mind, with some reservations to be noticed immediately, are accessible to direct observation. But this direct observation is permitted to us only as individuals. Hence that unique sense of ownership, with which we regard our thoughts and feelings; they are ours in a way which excludes all other persons from any immediate share in them. At the same time we are in like manner shut out from the inner lives of our fellows.

On this fact is based the important distinction between internal and external, or to slightly change the phrases, between mind and body. Whatever fact is beyond our immediate consciousness is external to us; hence even the body, which seems so near, is for the psychologist as much an external object as any other material object. And generally, all operations of mind, in whatever consciousness they present themselves, are marked off from physical, chemical, and biological processes, as the mental from the material.

The observation of our own states is called introspection, or looking within, to mark it off from that outward gaze upon surrounding objects which, though it is an indirect process performed by interpreting our impressions into their causes, is nevertheless the ordinary attitude of every one. In practice, however, difficulties turn up in applying this process of introspection, which somewhat diminish its value. By the nature of the case, only one observer can record what is going on in any particular mind; thus there is a risk of error for want of the check which is afforded by comparing independent results. In a matter, also, which so intimately concerns us as this inner life, it is unusually difficult to be candid and impartial, and so the process is vitiated not only by the want of a check upon individual observations, but also by the partiality of the only witness.

Again, the act of attention, which introspection involves, is one which cannot accompany any but very intellectual states of mind. We can take note, for example, of what is presented to consciousness in solving an algebraical problem, but few men are so imbued with the scientific spirit as calmly to register their feelings, when they pass through some moving experience, some agony of pain or pleasure. Shut off in such cases from immediate observation at the moment, we must perforce content ourselves with looking back, when the states to be observed have passed away, and when our results have all the imperfections of a remembered, as opposed to a direct, observation.

Nevertheless, with due precaution, this instrument of introspection will prove valuable, and the reader will often be called upon to exercise it in the sequel. The account of mental phenomena, which will be laid before him, is offered for comparison with his own inner experience, and any cogency which it may possess will be due to its correspondence with that experience.

5. **The Instruments of Psychology: External Observation.**—Opposed to the internal process of observation stands the external one. By this we observe the mental phenomena of other beings than ourselves. Since this cannot be done directly, it must be done indirectly, that is, through those evidences of mental operations which can be observed by us. In ordinary life, language and gesture and written or printed signs are usually available for this

purpose. Side by side with these intentional evidences, come unintentional ones which are not less instructive. Custom, in its various dresses of religious observance law fashion and art fine and not fine, may be made to yield very valuable results in the hands of a skilful psychologist. Of course the records of individual experience in books of travel, correspondence, diaries, and autobiographies, offer useful material. Of late, indeed, a powerful artistic tendency has come into play with the avowed aim of recording impressions and experiences on their inner, or subjective, side, in preference to their external side. In painting, the impressionists show how things would seem to you if you didn't know what they were. In literature, places situations and characters are tasted, so to speak, like tea by a tea-blender, and the practice is sometimes continued over a whole life, the results being gathered together in a journal. It would be a little rash, however, to take such observations as exactly true. When a diary is kept, an important consideration is what ought to be felt, no less than what is felt, and the two considerations do not always coincide in their effects. The alternate protestations of Marie Bashkirtseff that her journal was a photograph of her thoughts, and that it was an idealization of them, point to the two features of such records which must always be kept in view. Now she asks: "Why tell lies and play a part?" and assures her readers that she reveals herself completely. Soon after she warns against a too ready belief: "I did not think of A—— as I wrote. . . . I idealized him to make it more like a romance." This observation of individuals by themselves has been undertaken further for a scientific purpose by the blind, the deaf and dumb, and the epileptic. Observations have also been made upon the earliest stages of child-life by writers like Preyer, Darwin, and Taine. Somewhat similarly criminals, insane persons, and the subjects of hypnotic experiments have been observed for the purposes of psychology, and observations made by travellers upon savage races are a fruitful source of illustration and suggestion.

6. **The Instruments of Psychology: Experiment.**—Such, in brief, are the two time-honoured instruments of psychology, subjective and objective observation. Lately they have been supplemented by a third, experiment, and psychology is being converted into an experimental science. The acknowledged sacredness of the human personality will always tend to limit the scope of experiment to those mental phenomena which may be studied without damage to the physical organism, and without risking moral deterioration. The practice of vivisection has arisen in cases where danger of the first kind exists, as, for example, when it has been desired to trace out the functions of the various nervous structures; and thus apes, dogs, and rabbits have found how perilous it is to exist as Things in a world of Persons. But scientific zeal is not always to be restrained even within these limits. "Bartholow, an American doctor, with a disregard for the human subject to which we have not attained in Europe, has reproduced on a man, whose brain was exposed by a gunshot wound, the experiments which Ferrier made on dogs" (Lagrange). The practice of hypnotizing persons for public entertainment, as at present performed by individuals without medical training and without a due sense of responsibility, is to be condemned both for the physical and the moral hurt to which its subjects are exposed, and is in reality not less obnoxious than cases of the kind just mentioned.

The experiments used to determine the relation of sensation to sensory stimulus, and similar experiments, do not fall under this ban if performed with due care; they have already yielded a harvest even less notable in itself than as indicating the gradual transformation of psychology into an exact science. The observations of human faculty, which are being taken in anthropometrical laboratories, may be expected to lead to the determination of many important constants or units of measurement; for, of course, in all measurements one magnitude is compared with another; numbers have no meaning unless we know what they stand for; and in order that the measurements taken by one person may be understood by another, they must be expressed in terms which are known to this other. One of the chief requirements, therefore, of the psychologist is to have laid down the quality and amount of faculty which he may regard as the average, and by which, as a standard, he may measure any abnormalities. Following out this idea, psychological laboratories have been founded in various places—Leipsic, Berlin, Cambridge, Harvard, and elsewhere. Wundt led the way in founding the Leipsic laboratory, and it may be taken as a type of the rest. "In 1879 rooms for the laboratory were set apart in the university buildings, the authorities also granted a yearly appropriation for the purchase of apparatus, and more recently a demonstrator with a salary has been appointed. . . . The students come from all quarters—it should be added, except from England—there are nearly always Americans and Russians, and often Scandinavians, Czechs, Greeks, and Frenchmen. The men work in groups, at least two are needed to carry on most psychological experiments, the one acting as subject, the other taking charge of the apparatus and registering the results. The chief inquiries pursued have been concerned with the analysis and measurement of sensation, the duration of mental processes, the time-sense, attention, memory, and the association of ideas."[1]

7. **Descriptive Psychology.**—Such being the nature of the subject-matter, and such the instruments of psychology, this chapter will be appropriately concluded by considering what methods promise the best results. We might expect that the methods employed by natural history, which is the science most closely allied to psychology, would offer fruitful hints. The student of natural history has a double task. On the one hand he has to trace out the different groups in which animals may be arranged—here his work is descriptive; on the other hand, he has to investigate the laws of their development, and for this purpose employs the inductive method. In a science like psychology, which also deals with phenomena of life, though of a very special kind, we may follow with advantage the method of natural history, and divide our treatment of the subject into descriptive and inductive. The first step, then, will be to enumerate the various species of conscious states. This process has already been roughly performed for themselves by the great mass of people in close connection with the theories of writers who happen to have caught their ear, and who usually belong to the age just past. The result is the familiar classification of mental facts into sensation, memory, imagination, judgment, feeling, emotion, purpose, desire, and so forth. Popular psychology not only owes much to the commonplaces of literature on the subject, it is also indebted to the purveyors of phrenological and other pseudo-scientific wares, who occupy in the study of mind a place corresponding to that of the vendors of quack drugs

---

[1] *Mind*, xiii. 37.

in medicine. Conclusions so attained stand in need of the closest scrutiny, and indeed this task has always been regarded as preliminary by the expounders of new systems. The ideal to be followed is the grouping of mental phenomena under as few heads as is consistent with their different forms. An important step in this direction was that of the German, Mendelssohn, who made the triple division into thought, feeling, and will. Since his time various attempts have been made to reduce these principles to a single principle. The ways in which this has been done are very various. The most striking is, perhaps, the theory of Herbart, who regards all mental phenomena as forces which react on one another, and are therefore measured by their strength and not by any difference in quality.

The fact, however, that mental phenomena never occur singly, but always in some complex combination or other, stands in the way of a correct classification. Thought, feeling, and will do not lie side by side, as it were, like stones in a mosaic, any of which could be removed without destroying the rest; they rather resemble the functions of the body, none of which are possible without the co-operation of all the others. A man's opinions, for example, are due not merely to his having reasoned them out, but more than he himself or others imagine, to favourite associations or personal pique. We should not be justified, therefore, in classifying his opinions as purely intellectual; they also include emotional elements.

8. Cause and Effect in Psychology.[1]—The causes which give rise to states of mind may be divided into states of mind and states of body. Thus the sensation of red is due to a light-stimulus received from some object by way of the

---

[1] See for this paragraph *Mill's Logic*, Book iii., cc. 8 and 10.

eye. Here a state of mind, the sensation of red, is caused by a state of body, namely, a change in the retina. All cases of this kind, in which a mental effect follows upon a physical cause, are instances of psycho-physical laws.

On the other hand, states of mind are very often due to previous states of mind, as when to us being in a reverie, one thought suggests another. Here a state of mind, the thought suggested, is caused by another state of mind, the thought which suggests. Cases like this, in which a mental effect follows upon a mental cause, are instances of mental laws.

In determining how particular states of mind give rise to one another, we are met by the same difficulty as was noticed with respect to classification. Thoughts, feelings, and purposes are entangled together almost inextricably. Thus, suppose that we were investigating the laws under which the sentiment of sympathy was generated, it would be impossible to produce, for the purpose of our investigation, a state of sympathy without at the same time heightening intellectual processes, and we could never be quite sure that these were not the cause of the sympathy. We feel deeply with others, when we have wide thoughts. Even assuming, therefore, that the fullest freedom of experiment were allowed, which, as a matter of fact, is almost inconceivable, all our results would be rendered faulty by being unable to vary one circumstance at a time. In respect, therefore, of the more complex states of mind, we may not hope to obtain such certainty as when Mill's Method of Difference can be employed. It is never possible in the study of mind, on its higher side, to introduce a single new factor into a set of circumstances which are perfectly known, and then to observe the difference which is caused by it. For the study of mind is made difficult not only by the complex conditions of each mental state, but also by the way in which the several effects of those conditions are blended, transformed, annulled in the state of mind to which they give rise. The consciousness of duty performed blends with the consciousness of healthy physical existence into the single feeling of well-being. We have to contend with a complication of the effects, as well as with a plurality of the causes. Each change in a concept, an emotion, a purpose, acts and reacts upon all the other elements of the same state of mind. We often find that a cherished ideal surrendered, or a new purpose formed, puts all the objects of our mental landscape in a new perspective.

This difficulty only holds with respect to those cases which have been distinguished above as instances of mental laws. Where, however, a sensation is produced by some external cause, the problem is much easier. Many converging associations point to the fact that we see by means of the eye, hear with the ear, and so forth. We have to deal here with physical laws, and can proceed according to physical methods. There is a special reason, too, why the mental modification which follows upon a sensory impression can be marked off from the other elements of the same state of mind. A sensation is an element which has not yet been brought into relation with the other contents of consciousness; a sensation of red has no meaning until we connect it with other sensations; but it is then no longer a simple sensation, it is part of a perception. Thus, in the investigation of a sensation, we can turn the attention aside from all those accompanying elements which obstruct our view of the more complex states of mind.

We have proceeded hitherto as if states of mind were due either to states of body or to other states of mind. Usually, however, they are due to these two sets of causes acting together. Hence, in enumerating the conditions which give rise to any state of consciousness, account must be taken of both. All the physiological processes involved in the life of the body are factors, so far as they affect, even in the most indirect manner, the state of the mind, and few, if any, are without such influence. Hence they demand notice no less than those previous states of mind, which alone perhaps are observed by us. And so even external objects, which in any way produce changes in the states of the body, are factors in the life of the mind. They may be marked off as the remoter, from the nearer or physiological, conditions of consciousness.

Yet we may not think that mental facts are completely explained when we have enumerated these external conditions. As we advance from the more general to the less general sciences, we are continually confronted by some fresh quality of the facts with which we have to deal in. Thus all facts, so far as they relate to number and extension, are capable of geometrical treatment; but geometry cannot deal unaided with questions of force—these are left to mechanics. Again chemistry considers bodies not only as being of certain magnitudes, and as manifesting various kinds of force, it also regards them as differing in quality. A pound of lead and a pound of feathers are indifferent to a student of statics, but not to a chemist. We shall find this relation between the more general, and the less general, sciences to hold good when we pass from physiology to psychology. After making use of all the assistance afforded by mechanics, electricity, chemistry, and physiology in the explanation of psychical processes, and after making all legitimate use of analogies drawn from these branches of

c

science, there remains a residuum of facts that cannot be accounted for by considerations drawn from external sources. The generalizations to which we are led by induction from these special facts of mind, form the special laws of mind.

On the other hand, the facts of muscular movement show that mental factors co-operate in physiological changes. Hypnotism has given some striking confirmations of this truth. "Here," says Dr. Moll, "the bodily functions show a deviation from the normal, purely as a consequence of psychical states. Just as a man paralyzed by fright cannot move in consequence of a mental shock, and not in consequence of an injury to the muscles, so people in a state of religious excitement have visions, not because their eyes are abnormal in visual function, but because they are in an abnormal mental state; and in hypnosis (*i. e.* the state of being hypnotized), the muscles, the organs of sense, &c., are abnormal in function only because the mental state is altered."[1]

9. **Inductive Investigation.**—For reasons referred to in the last paragraph, the relation of a sense-stimulus to the resulting sensation presents a favourable opportunity for experiment. It is one of the few cases in the study of mind in which experimental verification is possible. It may therefore stand as a type of the procedure to which research should conform as far as possible. The law suspected by Weber that the stimulus applied to a sense-organ must be increased in geometrical progression in order that the resulting sensation may increase in arithmetical progression, has been followed out into its results, and these have been compared with the results obtained by experiment. The calculated and the observed results have been sufficiently close to justify a high degree of certainty being attributed

[1] *Hypnotism*, p. 53.

to the law; the deviations of the actual from the calculated results being due, in part at least, to the interference of other causes.

The other methods about to be enumerated suffer from imperfections of various kinds which render them much less reliable than this. But this drawback is not limited to their use in psychology. There is no ground, therefore, for a despondent attitude in face of the problems of the science; its resources are only just beginning to be explored.

In the past, experiment has rarely been attempted, and inquirers have been limited to observations under more or less unfavourable conditions. By observing the ways in which states of mind occur, together or in succession, empirical generalizations have indeed been set forth which embody the more frequent forms of such connection, as, for example, in the laws of association. But all these inferences have been vitiated by the inherent defects of the Method of Agreement. Because certain effects have invariably been preceded by certain conditions, it does not necessarily follow that these conditions were the causes of those effects; both may have been due to some circumstance which passed unobserved. Thus, the outbursts of anger, or other violent passion, which frequently precede an attack of insanity, are not to be regarded as its cause. They are rather earlier symptoms of a diseased nervous system of which the insanity is merely a later symptom. And generally, the danger of inferring from psychological observations which are unchecked by experiment consists in this, that earlier symptoms may be mistaken for the causes of the later symptoms, whereas, in truth, both earlier and later symptoms are caused by the same circumstance.

The Method of Variations has been applied with some

success to statistics, and results which are plausible, and may be true, have been obtained. Thus the relation between physical and social conditions on the one hand, and nervous disease and crime on the other, has been investigated. Morselli's book on *Suicide* is the record of an attempt in this direction. It is obvious that success depends on the figures compared being of the same denomination, that is, denoting the same classes of facts. And it is sometimes forgotten that mental phenomena cannot be referred to any single set of causes however important they may be.

This method has also been used to unravel the complexity of some of the higher states of consciousness. True we may not regard any side of the mind to the exclusion of the rest. We may not consider the emotions alone and apart from the intellectual processes with which they are so closely connected. But we can look for cases in which they are comparatively undeveloped, and again for cases in which they are very highly developed, and note the difference in the conditions which give rise to the difference in the results. Much of the advance which has marked the recent history of psychology has been made in this way. The observation of the acts of young children and of animals affords inferences as to the nature of comparatively simple and undeveloped mental states. The observation of genius, of mental derangement, and of crime will furnish cases in which some functions are excessively developed, and others in which they are hardly developed at all. Thus, taking the average adult mind as the standard, we can measure off differences in two directions: now in a negative direction towards the simpler forms, now in a positive direction towards the cases of very high or very distorted development.

CHAPTER II.

GENERAL CONDITIONS OF MENTAL DEVELOPMENT.

10. **The Elements of the Nervous System.**—A man is walking along the street in full enjoyment of conscious life. His foot slips on a piece of orange-peel, and falling heavily, he strikes his head on the sharp edge of the granite kerb. He is carried home in an unconscious state, and does not come to himself again, until, by a skilful operation, his brain is relieved of the pressure caused by his fractured skull. Take another fact: an application of arsenic destroys the tissues of the nerve of a tooth, and the pain of having it stopped is lessened, if not done away with altogether. By considerations like these, we are led to connect the changes in our consciousness with the functions of a particular set of organs, those, namely, which form the nervous system.

Roughly speaking, nervous structures may be classified into combinations of nerve-fibres and nerve-cells. By the combination of these cells and fibres in various ways, the whole nervous system is built up; and the general principle of their distribution is that the nerve-cells are grouped together at various points in the body, while the fibres connect the groups so formed. What we usually call a nerve, is really composed of several bundles of nerve-fibres bound together by sheaths of connective tissue; the nerve-fibres themselves are one thousandth part of an inch or less in diameter. Each nerve-fibre has a distinct course, and is insulated from other nerve-fibres.

The groups into which nerve-cells are collected, are called nerve-centres, since they form the points to and from which the nerve-fibres run. They contain some fibres in addition to the cells of which they are chiefly built up. The nerve-bundles, on the other hand, are wholly composed of fibres. And so while the fibres extend over all the parts of the nervous system, the cells are confined to certain parts. "Such nerve-cells are found in various parts of the brain and spinal cord, in the sympathetic ganglia, and also in the ganglia belonging to spinal nerves as well as in certain sensory organs, such as the retina and the internal ear."[1]

The processes which take place in the cells and fibres are called nervous excitations. They arise when some disturbing agent is applied to a nervous structure; this may be a ray of light, a sound, a moving solid body, and so forth. Such a disturbing agent is called a *stimulus*, and the act of disturbance is called stimulation. These nervous processes are distinguished according to the direction in which they are transmitted. Sometimes they begin at the surface of the body and pass on to the central nervous organs; the impressions received by way of the sense-organs, such as the eye or the skin, are examples of this class. Sometimes the excitation begins in the central parts and travels to the surface, as, for instance, when an impulse from some nervous centre causes a muscle to contract. The processes which go to the centre, are often called centripetal, or sensory, excitations; those which come from

[1] Huxley, *Physiology*, p. 279.

the centre, are called centrifugal, or motor, excitations. The names, *centripetal*, *centrifugal*, refer to the directions in which the excitations travel; while the names, *sensory*, and *motor*, refer to the effects produced by them. These several names are also given, by a very natural metaphor, to the nerves which convey the excitations.

The sensory nerves are provided at their outer terminations with organs of such a nature as to be stimulated by special agents: thus the retina is susceptible to light waves, the ear to sound waves. Ordinarily the form of these organs protects them from any influences other than the accustomed ones. If, however, by some accident another agent is applied to them, the result is the same as would be caused by a stimulus of the usual kind. If, for instance, the finger be pressed against the outer edge of the eyeball, an illuminated figure appears on the inner side of the field of vision. In other words, although the agent is not a ray of light, the impression is still a visual one, that is, a sensation of light.

On the strength of facts like this, a so-called *specific energy* has been attributed to the nerves; by this term is meant that each nerve, however stimulated, is capable of causing only certain kinds of impressions. Thus, it has been said, the optic nerve, however stimulated, can only give rise to sensations of vision; the nerves of smell can only give rise to sensations of smell; and this because of some peculiarity in their structure by which each kind of nerve can give rise only to certain kinds of impressions. But no corresponding differences can be detected in the structures of these nerves—nerves that bring light impressions are indistinguishable from nerves that bring touch impressions. Hence the differences which we are conscious of in the impressions brought by various nerves must be due to differences in the structures which form either their outer terminations at the surface, or their inner terminations at the brain; and, indeed, the very diverse forms of these terminations make it conceivable why this difference of function arises. Each of them seems adapted to transform all the stimuli which reach them into excitations of some special kind. The retina, for example, may be supposed to transform all sorts of stimuli into sight impressions; and so generally, the manner in which a nerve will act, depends upon the structures at its terminations.

The outer, or surface, termination of a motor nerve is connected with a muscle, and the effect of an excitation passing along the nerve is to cause the muscle to contract. All the movements of the body, various and complex as they may be, are caused by this simple means, that is, by the contractions of the muscles, singly or in combination. Motor nerves cannot be distinguished in structure from sensory nerves; they are provided, however, with end-structures of such a kind, that only the excitations which travel from the centre can produce any effect. "A motor nerve is connected with nothing which can make a change apparent save a muscle, and a sensory nerve with nothing which can show an effect but the central nervous system."[1]

11. **The Nervous System as a Whole.**—We now proceed to regard the nervous system as a whole. It consists of two sets of nerves and nerve-centres, which are intimately connected together, and yet may be studied apart. These are the cerebro-spinal system, and the sympathetic system. As far as is known at present, the latter is not immediately connected with any psychical processes, and may be left

---

[1] Huxley, *Physiology*, p. 284.

out of account. The cerebro-spinal system with which we are more closely concerned, consists of the cerebro-spinal axis (composed of the brain and spinal cord) and the cranial and spinal nerves, which are connected with this axis.

The spinal cord lies in the cavity of the spinal column. From it there come thirty-one pairs of spinal nerves which are attached to it as follows: Each nerve, on leaving the column, is double for a short distance, or to put the same fact in another way, the bundle of fibres of which each nerve is composed, is separated into two smaller bundles before it enters the spinal cord. That root which lies nearer to the front of the column is called anterior; that root which lies nearer to the back is called posterior. If a spinal nerve is irritated before its fibres separate into the two bundles, a double effect is observed. On the one hand, a sensation, generally painful, is referred to those parts of the body from which the nerve comes, and at the same time muscular contractions arise in the same parts. If, however, only the anterior root of such a nerve is irritated, muscular contractions arise but no sensations; while the irritation of a posterior root gives rise to sensation, but not to muscular contraction. The inference which may be drawn from these facts is that the fibres which enter the spinal column in front convey only motor impulses, while the posterior set convey only sensory excitations.

The function of the spinal nerves is to convey sensory excitations from the skin of the trunk and limbs to the nerve-centres in the spinal cord, and to send motor impulses back again from the spinal cord to the muscles of the same parts of the body.

The remainder of the cerebro-spinal axis is constituted by the medulla oblongata, the cerebellum, a number of smaller centres, and lastly, the great cerebral hemispheres which form the largest and most distinctive part of the human brain. While the centres of the spinal cord are arranged in a series, these higher centres are collected together in the skull, as though the end of the spinal cord had been folded over on itself like the end of a shepherd's crook.

The cranial nerves take their origin in these higher centres. There are twelve pairs of them, which are distributed as follows:—

The first and second pairs, counting from the top, are the olfactory and optic nerves. These are really continuations of the brain, and scarcely deserve the name of nerves. The olfactory nerves convey impressions of smell, while the optic nerves convey those of sight. The remaining ten pairs are attached to the medulla oblongata. Taking up the enumeration, the third, fourth, and sixth pairs are attached to the muscles of the eye. The fifth pair go to the muscles of the jaws, and the skin of the face; they also convey taste impressions from the mucous membrane of the tongue. The seventh pair are connected with the muscles of the face, and are called facial. The eighth pair are the auditory nerves, or nerves of hearing. The ninth pair are partly sensory, partly motor; they bring taste impressions from the hinder portions of the tongue, and govern the muscles of the pharynx. The tenth pair are interesting as the only pair of cranial nerves which supply those regions of the body which are distant from the head; they are connected with the lungs, the heart, the liver, and the stomach. The eleventh pair are motor nerves, which supply the muscles of the neck. The twelfth and last pair supply the muscles of the tongue.

The medulla oblongata is a continuation of the spinal cord, and closely resembles it in structure; it is chiefly distinguished from it by the occurrence of cellular masses, or ganglia.

The cerebellum is the first part of the brain which shows great divergence from the structure of the spinal cord. The gray cellular masses form an outer covering, which is clearly distinguished from the radiating white fibres which occupy the interior.

The cerebral hemispheres lie like a mantle over the other parts of the brain, and fill the whole upper portion of the skull. They are separated from each other by a deep cleft or fissure in the middle, and are only connected by a thick band of transverse fibres at the bottom of this fissure. Their outer surface, or *cortex*, is formed of folds of gray cellular matter, like that already described as being on the surface of the cerebellum. The interior of the hemispheres is composed of white fibres; these connect the various parts of the hemispheres with each other, and with other parts of the brain and spinal cord.

12. **The Functions of the Nervous System, I.**—When it has been once allowed that states of consciousness are connected with states of the nervous system, the next step will be to ascertain, if possible, what functions of the nervous system correspond to the several operations of consciousness. Stating the problem as it appears to the physiologist: What paths are taken by nervous excitations on their way to and from the highest centres? What part of the brain, for instance, controls the movements of the legs?

A very short experience is sufficient to connect the sensations of sight with the eye, those of touch with the skin, and so forth. The processes going on in the outer parts of the nervous system do not so much need explanation as those central processes which have their seat in the higher centres. These, as yet, are very imperfectly understood. Any inferences, therefore, that may be drawn with respect to them, must be taken as merely tentative.

There are three ways in which this question may be approached; namely, physiological experiment, anatomical investigation, pathological observation. Their nature will be explained in the succeeding paragraphs.[1]

Physiological experiment has, for its end, to determine the functions of particular nervous structures, and for the reasons which were pointed out in § 6, is not usually possible on the human subject. Hence animals which possess a nervous system analogous to that of man, have been made use of. The practice of experimenting on the living animal has been introduced, because the continued activity of the higher nervous centres requires the co-operation of the other physiological functions of the animal, and this is only possible of course in the living creature. It is obvious from this that the practice of vivisection can only be justified upon two conditions: First, that the experiments are reasonably likely to afford important results; secondly, that they are undertaken with all possible precautions for the avoidance of unnecessary pain. The disrepute into which the practice has fallen in some quarters, and not undeservedly, has been due to the neglect of these conditions. This is not the place, however, to enter further into this vexed question. The experiments which have been performed may be ranged under two heads. On the one hand, stimuli have been applied to various parts of the brain, and the

[1] Wundt, *Physiologische Psychologie*, vol. i. p. 98.

muscular contractions following thereupon have been observed; or, on the other hand, portions of the brain have been removed in order to observe what functions are thereby interrupted or disturbed, the inference being that these interrupted or disturbed functions have their seat in the parts affected by the experiment. But these methods are subject to very serious limitations. In the first place, it is almost impossible to excite a portion of the brain even with the help of the most delicate stimulus without throwing the neighbouring parts into agitation of a more or less serious kind, while the experiments which involve the destruction or removal of nervous tissues run the risk of causing irritation in the parts which remain; and we may mistake the effects of this irritation for those which are simply due to the cessation of the functions of the parts removed. A further difficulty lies in that property of the nervous system, by which the function of an organ, which is diseased or lost, may be gradually taken over by some other part of the system; in this way the effects of such a removal are compensated and concealed.

Anatomical investigation attempts to trace out the paths taken by nervous fibres, and therefore, presumably, by the excitations which they transmit. But this method again can help us only a very little. The complicated structure of the higher centres makes it almost impossible to trace any single set of fibres very far without losing sight of them. Much less is it possible to follow any single fibre from its outer termination at the skin all the way to the highest centre of all, the cerebrum. On the other hand, comparative anatomy has furnished many valuable clues to the functions of the various parts of the brain and the connections between them.

Pathological observation, in the third place, has for its field diseased states of the brain. By observing the resulting interferences with muscular contractions, with sensations of various kinds, and with the higher mental operations, inferences can be drawn as to the way in which particular nervous centres are connected with particular mental functions. The fact that disuse of an organ is often accompanied by degeneration of the related parts of the brain, is of importance here; the loss of sight or hearing in early life is often followed by atrophy of certain parts of the brain. Changes in the structure of the brain, of which nervous diseases are the outward expression, are a further source of evidence. All this evidence, however, is to be taken with the same limitations as those pointed out under the head of physiological experiment.

13. **The Functions of the Nervous System, II.**—The simplest functions of the central parts of the nervous system consist in transforming the sensory excitations received from the surface of the body into motor impulses. This involves a reflection, or "bending back" so to speak, of the incoming process into an outgoing process; hence a sensory impression taken along with the connected motor impulse is called a *reflex*. If the foot of a sleeper is touched with a feather, the limb will be drawn up. Notice what is involved; the sensory excitation caused by the feather is transmitted by a centripetal nerve to a central part of the nervous system (in this case, the spinal cord), where it is reflected, or bent back, in the shape of a motor impulse; and this motor impulse causes the contractions in the muscles by which the foot is drawn up. In proportion as the stimulus is stronger, the resulting movements are more extensive and more vigorous. The start which we give on

touching some red-hot object, is an example of a reflex action which is spread over a large number of muscles. At the other end of the scale, very faint stimuli do not set up any reflex actions whatever; for example, the faint air-currents which are always playing upon the surface of the body.

This transformation of sensory excitation into motor impulses is the business of the nervous centres. But the lower centres need not necessarily receive a stimulus from a sensory nerve in order to fall into activity and send forth a motor impulse: an impulse coming from a higher centre will have the same effect. The sleeper on waking will move his foot, not necessarily because some object has touched it, but in obedience to an impulse from the brain.

There is a kind of hierarchy among the various centres, and the lower are placed in subordination under the higher. The nerves, for instance, which lead to the various sets of muscles are grouped together at various points before they reach the higher centres, in such a way that an impulse coming from a higher centre to a lower one will set all the muscles to contract which are governed from this lower centre. A very slight impulse may thus suffice to cause a relatively violent and extensive contraction of the muscles. Instead of special motor impulses passing from the higher centre to each muscle concerned in the act of walking, it is sufficient if the centres are excited which govern the motor nerves going to the muscles in question; just as a general gives his orders to the officer at the head of a regiment instead of to every individual soldier.

In the healthy individual, impulses are always flowing from the higher centres to the lower ones, checking the activities into which they would otherwise fall amid the constant stimuli which are produced by surrounding objects or by the changing states of the body. Thus the control of the lower centres by the higher has its restraining, as well as its impelling, side.

There is also a reaction of the lower centres upon the higher. This reaction comes chiefly from the centres to which sensory nerves converge; the impressions received from the surface of the body furnish a stimulus to the higher centres without which they degenerate for want of activity. And so, looking at mental processes in their mechanical aspect, they seem to consist in a stream of energy which flows from the sensory organs at the surface to the central organs bracing up the latter and being by them transformed into a returning stream; this acts upon the lower motor centres to excite or to control, and so finally returns by way of muscular activities to the outer play of forces in which it took its origin.

Let us now proceed to trace out the functions of these nervous centres, beginning with the lowest, and observing how they become more and more complex until we reach the highest.

The spinal cord is the place where sensory excitations, which have been received from the skin of the trunk and limbs, are transformed into motor impulses. The spinal cord seems to be under the continual control of the brain, for when that has been removed, the excitability of the spinal cord is increased.

The reflex actions which are executed by means of the medulla oblongata, are of a much more complex nature than those performed by means of the spinal cord. They include the movements of in- and expiration, swallowing, and those concerned in the circulation of the blood. It is not surprising that a nerve-centre which stands in the manifold relations to the outer world indicated by the list of nerves given in § 11, should also control the means by which the resources of the body are husbanded. The reflex actions of the medulla oblongata are thus marked off from those of the spinal cord by their greater purposefulness, a kind of self-regulation being brought about by their means with respect to nutrition and physiological waste.

The functions of the cerebellum are connected with the combination together of the various muscular actions. Hence, for example, it is found very highly developed in creatures such as birds, which have to perform very complex and rapid movements. When it is the seat of disease, the control of the muscles is disturbed; the patient moves unsteadily, and is affected with that feeling of giddiness which is the mental expression of this imperfect control. The unsteady steps of a man under the influence of alcohol may be explained by the effect exercised on the cerebellum by alcoholic liquor.

Turning now to the cerebral hemispheres, the most striking fact, perhaps, in relation to them is that either seems fitted to undertake alone nearly all the work which is usually done by the two in combination; and Dr. Maudsley has suggested that the processes of thought, apparently single as they are, in reality are double, and are blended together in the same way as the impressions made on the retinas of the two eyes fall into a single picture.[1] That there is a certain independence of either hemisphere from the other is proved by the following considerations. The cerebral hemispheres may singly suffer considerable disorder before any noticeable disturbance of nervous function takes place. Provided the disease is

[1] *Mind*, xiv. 161.

confined to one hemisphere (and cases are on record in which one whole hemisphere has been affected without completely deranging the mental life), the patient's general intelligence may be scarcely affected. Where, however, both hemispheres are diseased, or imperfectly developed, the subject is usually idiotic. He can perform the simpler actions, indeed, but is unable to adapt himself to the continual changes in his surroundings. "If left to himself, he will set himself on fire, or fall into the water, or cut himself, or get entangled in a machine, or come to some actual physical harm which could have been avoided by the exercise of rudimentary intelligence."[1] For the control exercised by these, the highest centres, upon the lower, involves the adjustment of the functions of the latter to the more or less intricate trains of events which go on around us.

This control is apparently exercised by certain parts of the cerebral rind or cortex. For example, the centres which control the impulses to the muscles are grouped together in the middle portion of the external surface of each hemisphere. The sensory centres, in which are registered the effects of the impressions successively received by way of the sense organs, probably occupy the regions behind this motor zone. One of the best established results is the location of the speech-centre in the base of the third frontal convolution of the left hemisphere.

How are we to think of the operations of this highest element in the nervous mechanism? Since, especially, it has to bring into relation all the sensory activities on the one hand, with all the muscular activities on the other, each and all of these must be represented in it. But the representation need not be a direct one; it will be sufficient, if

[1] Mercier, *Sanity and Insanity*, p. 291.

each combination of a set of motor impulses with the answering set of sensory impressions (that is to say, each reflex), which is carried out by means of the lower centres, leaves its mark in these, the highest centres. In other words, the processes which go on in the cerebral hemispheres are, so to speak, symbolic of those which have taken place in the lower centres. The continual interaction with our surroundings which makes up life will have for its effect the formation of new combinations of sensory with motor excitations in the lower centres, new reflexes in fact, and side by side, an answering series of representative marks in the highest centre. Take piano-playing, for instance; a practised performer can play a piece of familiar music from beginning to end with scarcely a look at the notes, or a thought as to his fingers. The sensations of touch and movement produced by playing each chord call up by reflex action the movements necessary to play the next chord. If the piece is slightly less familiar, one or two references to the notes may be required. The successive motor impulses to the arms and fingers of the performer are not yet so conjoined by practice with the successive impressions of touch and movement received from them that the motor impulses immediately follow upon the sensory impressions. Continual exercise, however, renders each train of movements which at first requires for its execution the co-operation of conscious purpose, gradually more and more automatic. And when the connection of each movement with the next is accomplished, the whole series of movements will follow automatically on the first; for this will immediately call up the second and so on to the end. Any one of the earlier movements will thus necessarily be followed by all the later ones, and may represent them in the supreme region of the nervous system. The first two or three members of a series are generally so used: the first two or three notes of a melody, say, *Call me back again*, or *Where is my boy to-night?* at once indicate the kind of torture the barrel-organ in the next street is about to inflict. We may thus conceive the cortex to be occupied by activities of this symbolic kind, each of them tending to call up a whole series of related activities of the lower centres. Words, whether read, heard, or spoken, be it in fact or in idea, have this representative quality to a special extent. Whatever form the symbol of a word may take, it stands for and may recall at need the intermingled series of functions which accompany the pronunciation of a word, namely, the excitations of the nerves of hearing, of the nerves of touch and muscular sense in the mouth, and the motor impulses to the muscles which are employed in voice-production. Further, a word usually implies a more or less complex train of associations which constitutes its meaning. It is this property of words, their standing, so to speak, at the head of a succession of mental operations following their suggestion like the genii of the lamp, which makes them the counters of abstract thought.

The process of conception may be illustrated by the same analogy; it is the capacity possessed by suitable symbols of calling up the reflexes (combinations of sensory excitations with motor impulses) which are involved in the meaning of the concept. Thus the concept of a house implies that certain sensory impressions of form and colour are connected with certain movements, such as walking up to and round the building. And notice that the appearance of a house from a particular point of view is associated with those movements by which the observer reached that point of view, and that the thought of the impressions and of the answering movements cannot be separated. When we think of some impression, there comes to mind by a kind of necessity the idea of ourselves as receiving that impression. A reflex in idea is the thought of ourselves as feeling something and doing something at the same time. Each element of thought, then, has its active no less than its receptive side. Hence, it has been paradoxically said, we cannot imagine the appearance of things which lie behind us; we always face round in idea to look at them.

14. **Cumulative effects of Interaction with Environment.**—The human being is constantly under the influence of surrounding forces, and each impression leaves a permanent mark upon him, which modifies his attitude to all future influences. Nor do these changes compensate one another, and, after a cycle of variations, bring him once more to the state from which they caused him to diverge. Insignificant most of them taken singly, they suffice in combination to effect a thorough-going transformation of his nature. His interaction with the objects in his environment changes him within the space of a few years from a helpless infant to the boy in full possession of his faculties, then in turn to the man in the prime of life, then to the old man, and lastly resolves his body into the elements out of which it was constituted. Side by side with this physical development, there goes a mental one not less striking in the variety of its several stages. The unshaped thoughts and undisciplined impulses of the child are transformed bit by bit into the comparatively controlled action and the clearer mental vision of the adult, and finally into the quieter thoughts and the love of peace, which distinguish more advanced years. With the greatest variety

in the details of this development, there goes a certain order in its stages. Each is due to the reaction of the preceding one upon environing circumstances, and so the whole series is, in a sense, involved in the first stage. The nature of the child is implicitly what later he will be explicitly. We shall now consider the chief conditions under which this nature unfolds itself; we shall take note of the physical conditions of that mental development with which we are more particularly concerned, so far as they throw light upon it.

15. **The Form of Mental Development.**—Development in amount, that is to say, growth, is the most obvious form. We are familiar enough with the gradual increase of the organism to full size. The accumulation of facts in the memory illustrates an analogous mental process.

Development in quality implies that the relations between the various parts of an organism become more definite, and more complex; sometimes this involves an actual transformation. The changes from larva to chrysalis, and then to moth, afford striking examples of development in structure. The building up of disconnected thoughts into concepts, involving as it does the gradual grasping of the ways in which these thoughts stand one to another, illustrates the way in which relations are formed between the elements of conscious life. We may expect to find that the reorganization of mental elements may at times produce results as unlike the causes from which they sprung, as the moth is unlike the chrysalis. J. S. Mill illustrated this metamorphosis by the chemical combination of oxygen and hydrogen gases into the liquid water; and opposed this mental process to those mentioned in the last paragraph in the same way as chemical to mechanical combinations.[1]

[1] *Logic*, Book vi. chap. iv. p. 3.

CONDITIONS OF MENTAL DEVELOPMENT. 39

States of mind are not always built up by the mere juxtaposition of their elements; it sometimes happens that the combination is quite unlike its elements (as water is unlike oxygen and hydrogen), and could not have been inferred from them.

Though development in amount, or growth, is usually accompanied by a parallel development in quality or structure, this is not always the case. The brain attains approximately its maximum size soon after the seventh year, but its structure is as yet comparatively undeveloped; the relations between its parts (the convolutions of the cerebral hemispheres, for instance), are as yet very imperfectly marked out. Similarly we may distinguish the mere memory for facts from the thorough understanding of the bearings they have one on another; one is mere learning, the other is scholarship, or scientific insight.

The development of an organism may be partial or general; it may be manifested in one or two organs, or in all. Similarly, in mind, one single faculty may exhibit remarkable excellence amid others of scarcely average quality. Such cases afford special opportunities for the study of particular faculties.

Even where the development takes a normal course, we can fix the attention on one side of it. Thus the changes in the brain may be studied apart from the other bodily functions. Similarly the memory and the will may be studied apart from other mental functions.

But it is only in thought that the processes of the physiological life may be separated from one another. In reality these are so connected together that each implies and is implied by the others. Respiration and the circulation of the blood are controlled by the brain, and cease if the controlling centres are seriously injured. On the other hand, the brain ceases to act if respiration and circulation are impeded. There is a similar correlation among the psychical processes. Each of the three species into which mental operations are classified—thought, feeling, and will—depend upon the two others. The following out of a train of thought by a schoolboy, as in the solution of an algebraical problem, is dependent upon the interest he takes in it; this is an emotional element. It also involves an active element, namely, the effort necessary to keep the attention fixed. A feeling—fear, for instance—rests on the act of thought by which the cause of the fear is brought before the mind, and excites us to flee from the cause; here, then, we have intellectual and active elements. Lastly, an act of will is determined by the influence which the end proposed has upon our feelings, and also on the knowledge of the actions by which that end may be attained. This mutual dependence is recognized in the French saying: "Il a les défauts de ses qualités." The connection between lying and a wish to oblige is not at first sight clear, but it holds good nevertheless. For example, the Indians of the Amazon Valley, according to Mr. Wallace, will "never refuse to undertake what is asked them, even when they are unable or do not intend to perform it."[1]

This development of the mind as a whole rests on the simultaneous development of the several mental functions, whereby they are fitted to take their part in the whole system. As we have seen, the means by which development takes place is the repeated exercise of faculty in answer to impressions received from without. In this way actions which at first severely taxed the attention, and were

[1] *Travels on the Amazon*, chap. xvii.

slowly and imperfectly performed, are at a later stage performed automatically, and with ease and rapidity. This habituation of the mind to the ready execution of its various functions is simply the acquisition of a number of habits; by exercise the mind comes to think, feel, and act, in particular ways. The extraordinary susceptibility of the young child to external influences, and the consequent ease with which its nature may be moulded in various directions, makes possible its gradual attainment of the developed powers of the average civilized man.

A function can be improved by exercise only up to a certain pitch. When that is reached, further exercise may not only fail to cause any advance, but may actually give rise to deterioration. Trainers are well acquainted with this fact, and relax the exercises of athletes from time to time lest they become "stale." At the same time exercise may remove this limit of faculty to a somewhat higher stage.

The effect of use and disuse on the various parts of the body is very striking; different occupations produce characteristic modifications of the physique. Thus, taking the average, the legs of the sailors employed in the American Civil War were one-fifth of an inch longer, and their arms one inch shorter, than those of the soldiers.[1] The effects produced upon the mind by the several occupations are not less striking; the clergyman, the lawyer, the physician, have each their class peculiarities. And by the law of correlation secondary effects due to class-bias turn up in the most unexpected quarters. Dr. Holmes has well said that you cannot keep thought in water-tight compartments; in spite of all enclosures it will percolate through the joints, and reach distant places by strangely devious channels.

[1] Darwin, *Descent of Man*, part i., chap. ii.

Gradual though it be, mental development is none the less marked by turning points in its progress. From time to time there arise stresses of circumstance which do not permit the individual to continue in the path he had marked out for himself, or at any rate had been previously following. These mental stresses usually accompany two critical eras in the physiological life, namely, the disturbances which take place on the organism attaining its full complement of functions; and again, when the prime of life is exchanged for the quieter existence of old age. The entrance upon school life, the choice of a vocation, marriage with its responsibilities, and retirement from business, also form crises in the average career; and the mental energy, which may have sufficed to meet ordinary demands upon it, often fails in the presence of greater emergencies. Hence one important condition of proper development is the maintenance of reserve energies on which, if occasion arise, draughts may be drawn.

16. **The Factors of Mental Development.**—We have just considered the form which is taken by mental development; it remains to consider the factors which bring it about. These may be divided into two sets, under the names nature and education. On the one hand there are the dispositions to think, feel, and act in certain ways, which, taken together, constitute the *nature* of the individual. On the other hand are the influences exercised upon him by his surroundings, by which his nature is drawn out and realized, that is to say, *educated*.

17. **Plasticity: Variety of Natural Dispositions, I.**—Among the many excellences which account for the domination of man over other animals, none perhaps is more striking than the variety of circumstance to which he can

adjust himself. This power of adaptation is called *plasticity*. Apart from that foresight with which he can take precautions against events distant in space or time, and transforms his surroundings to suit himself, his body of itself can become inured to the greatest possible diversity of climate and circumstance. The Fuegians live without clothing in a country where, even at the summer solstice, snow falls every day on the hills, and rain and sleet in the valleys. At the other extreme we find that human existence is maintained in sultry districts like the Coromandel coast, where the heat is so intense for some parts of the year as to destroy vegetation, the temperature ranging in the sun from 100 to 110 degrees. The mind, too, no less than the body, can become used to circumstances, which at first sight seem to render its activity inconceivable. The conditions of existence among the poorest of the poor would at first weigh as heavily on the mind of the average well-to-do Englishman as those physical extremes on his body.

Connected with this power of variation in response to varying circumstances, is the diversity of natural endowment with which each of us starts the world. It is a commonplace how far this diversity extends with respect to physique. Height, proportionate length of limb, complexion of skin, and especially the form of the features, and the colour of hair and eyes, so combine to produce a constant variation of appearance, that amid the millions of our fellow-creatures we rarely confuse one with another. This variation in physical structure is accompanied by a not less striking one in mental capacity. Some minds exhibit an unusual responsiveness to their environment, and their development is rapid all round. This sensitiveness to external influences does not necessarily indicate a natural equipment above the average; it may sometimes be the symptom of a mind too highly strung, or else of one which soon reaches the limit of its capacities, and is therefore less promising than one of slower growth, just as animals of a low grade of development like birds and reptiles reach their full complement of faculties sooner than animals of a higher grade, the elephant and the horse, for example. Apart from causes of this description, a readiness to develop quickly is the sign of a nature well adapted to its surroundings, and possessed of well-balanced faculties. If, on the other hand, we follow cases of less perfect development through all the successive degrees of imperfection, we find at first that fullness of development is merely delayed; then cases are reached in which the individual is weak-minded, and barely equal to the ordinary draughts upon his energy. As this imperfection becomes more marked, we meet with imbecility, and lastly with idiocy, in which the development of the individual's mind has barely been carried past the earliest stages.

Just as mental development in general can be measured by reference to the ordinary standard, so we may distinguish variations above and below the average in the case of the several faculties. Take music for an illustration: on the one hand we have precocity like that of Mozart, or of the child-pianists, Hoffmann and Hegner. Then come a large number whose taste for music is developed enough to render them respectable performers. They are followed by the average concert-goer, and lastly by those persons who have no ear for music, and can barely recognize the national anthem. Capacity for mathematics, natural history, languages, athletic exercises, good-fellowship, business transactions, public life, differs in similar ways. Some of these capacities seem so to be connected together that excellence in one is usually accompanied by excellence in another of them. A clever designer is usually, also, an excellent draughtsman. At the same time, the imperfect development of one or two special capacities does not of necessity betoken a mind below the average. Many children who scarcely can keep up with their fellows in the very special subjects of a school curriculum, turn out brilliant successes amid more congenial surroundings. Mr. Darwin, and his cousin, Mr. Galton, failed to be attracted either by school-life or medical studies. Richter tells a tale of a great mathematician, who from incapacity for ordinary studies or business, continued a day-labourer for thirty-eight years. On the other hand, boys who have shown the greatest promise amid the artificial conditions of school-life, sometimes fail completely when they have to deal with the problems of business. Here their partial capacities found an environment which exactly suited them; there they were unavailable.

18. **Natural Dispositions, II.: Heredity.**—Sometimes unusual forms of development are referred to the habits and manner of life of one's ancestors; it is said that faculties exercised and improved in the life of the individual are transmitted in this more advanced state to his posterity, so that they start at a point of development further on than that from which he started. On this principle Mr. Herbert Spencer has deduced the more complex faculties, and especially the moral ones, from the registration of experiences in the nervous system, and the consequent transmission of their effects to future generations.[1] Thus, it is urged, the individual is impelled to act in particular ways and for particular ends not by his own experience merely,

[1] *Principles of Psychology*, part viii., chap. vi.

but also by dim motives drawn from an experience not his own.

But the question has been raised whether dispositions acquired in the life of the individual are passed on. The facts, it is said, can be explained without the help of this theory. According to it numberless generations have handed on to their successors a slightly bettered outfit of faculties. But observe: the variation between different members of the same community often exceeds the difference between the average civilized man, and the savage from whose condition he is believed to have emerged, and so, if through any circumstance those individuals who exhibit particular qualities in their most developed form have the best chance of surviving, a difference is obtained at once, which is at least equivalent to the sum of the differences produced by inheritance acting over long periods of time. Hence the influence of such special inheritance must be very slight compared with that of natural selection. But even in cases where traits apparently acquired in the parent's life, have reappeared in his offspring, it is not necessary to assume that they have been specially handed down. That the parent placed under certain circumstances has exhibited certain dispositions, makes it probable that his offspring, if placed under like circumstances, will do the same. It seems a not uncommon occurrence for the handwriting of father and son to be almost indistinguishable, but we need not therefore assume that the tendency to write in that particular way was specially inherited. Those peculiarities in the nervous and muscular organization of the parent which made it natural for him to write in just that way, will, if they appear in the son, make it natural for him to write in that way too. To justify the theory of inherited acquisitions it would be necessary to show that the child's nature was an exaggerated copy of that of the parent: as if "the soldier's child is born to be yet more a soldier, and the politician's to be still more a politician." And this theory also requires that injuries suffered by the parent should reappear in the offspring. The examples of such inheritance are so rare, however, that they fail to give any adequate support to the hypothesis. The theory of heredity when fully stated includes the reappearance of the characters of remoter ancestors, no less than of immediate ones; and it is quite possible that traits which at first sight seem directly due to the influence of dispositions acquired during the parent's lifetime, are the effects of reversion to the characteristics of remoter ancestry. If we neglect for the moment the probability that the lines of an individual's ancestry intersect one another, we find of course that the number of his ancestors doubles with each generation that we go back. By the alternating predominance of this or that strain scope is afforded for the occurrence of variations not less numerous and striking than those which are actually observed. The many instances in which a characteristic leaps a generation to appear in the third, proves the possibility of this reversion. "We never see an individual in a face of any stock we know, but a mosaic copy of a pattern with fragmentary tints from this or that ancestor" (Holmes). The result to which the discussion of this question seems to be tending, is that the transmission of dispositions in an improved state, due to the experience of the parent, is indeed a factor to be taken into account, but that at the same time its importance has been very much exaggerated. The interest of this matter for the psychologist lies in the way in which it affects theories as to the origin of the moral and other sentiments.

There is one important corollary of the theory of development which throws light on the order in which the faculties develop; the successive stages of the individual's life history answer with some necessary limitations to the order in which those dispositions appeared in the history of the race. The man of the more advanced degree of development has passed through stages at which men of a lower degree have stopped. The civilized man, when a child, exhibits for a time those dispositions which permanently characterize the adult savage. We may suppose that both the higher and the lower races start on the same path of development, but that individuals of the one reach the limit of their progress sooner than individuals of the other. Negro children, when placed under tuition, often develop more rapidly than white children to begin with, but they sooner reach a stage at which further instruction has little or no effect.

19. Educational Influences, I.: Physical Environment. —Turning now from the tendencies implanted in the individual, let us examine the external conditions which call out or repress them. The objects amid which human life is passed make themselves known to us by their reactions upon our bodies. Through these effects they contribute to, or impede, our continual well-being. Gravitation, electricity, heat, light-waves, sound-waves, are some of the forms taken by these influences. The exact proportion and manner in which they affect us, is determined not only by their own laws, but also by the distribution and amount of the natural agents in which they take their rise. From this point of view we may classify our surroundings into various groups of natural agents: astronomical, the sun,

CONDITIONS OF MENTAL DEVELOPMENT. 49

moon, and stars; geological, the nature of the soil and climate, and the distribution of minerals; botanical, the nature and distribution of surrounding flora; zoological, the nature and distribution of surrounding fauna. All these circumstances tend to determine the conditions of human life. "In the rarefied air of the high Andes more respiration is required than in the plains, and in fact tribes living there have the chest and lungs developed to an extraordinary size."[1] Among other physiological influences, those of animal and vegetable diet respectively, are very marked; and the neighbourhood of other human beings has physical effects not less important, as, for instance, in the communication of infectious diseases.

Along with these physiological effects, there go psychological effects, some of them directly, others indirectly, derived from them. To make an absurd supposition, if man were a nocturnal animal, instead of one that seeks his food by day, his mental characteristics would probably be entirely different. The damp wintry days which make up so much of the English year, tend to drive Englishmen within doors, and in this way, by turning attention to home life, have powerfully affected the English character. Or to take an instance of another kind, the geographical conformation of Greece, by which the country is divided into small secluded portions through the intervention of high mountains or arms of the sea, was an important factor in the development of the Greeks of classical times. It favoured that isolation by which variations in a particular direction may proceed without being checked by the neutralizing effects of intercrossing. And so we may explain in part that very special development of capacity which marked

[1] Tylor, *Anthropology*, p. 74.

many of the Greek tribes. This seclusion was not seriously affected by the large sea-board. Racial pride tended to prevent intermarriage with barbarians and foreigners, and indeed usually found expression in legal enactments. And so great purity of descent was combined with great opportunities of contact with foreign ideas.

The comparative abundance and the nature of food, be it animal or vegetable, determines in definite directions the development of mental dispositions. The arts by which animal food is secured, include the exercise of alertness, ingenuity, and endurance in the chase. Further the invention and manufacture of weapons demands considerable capacity. The case is quite altered in tropical districts where sufficient vegetable food lies close at hand.

20. **Educational Influences, II. : Social Environment.**—The most important factor in our environment is the presence of other human beings. This has such far-reaching implications in mental development that it is usually singled out for special treatment under the name of the social factor. At the same time, we must remember that the effects produced on the individual by his fellows are of the same kind as those to which reference was made in the last paragraph. They consist in certain reactions which may be described in terms of physics and other sciences. It is the meaning which is read into it that makes the grasp of a friend's hand other than the mere contact with a foreign body.

Cases occur from time to time in which human beings have been detached at a very early age from the companionship of their fellows. In 1800 a human being, half savage, who lived in the woods, slept on dried leaves, and fled at the approach of men, was found in the department of the Aveyron by some sportsmen. He had no voice, and seemed very defective in intelligence. He had probably been lost or exposed in the woods when very young. His case shows how much the ordinary civilized human being owes to social influences. In their simplest form they consist of the care and training supplied by the parent. The parent's indirect influence is scarcely less important than that which he intentionally exercises; for his actions, and those of elder persons generally, are watched and imitated by the child. As the circle of the child gradually widens from home to school, and then to business and the great world, the social influences operating upon him become more numerous and more complicated.

Let us glance for a moment at the forms taken by those influences, and at their effects in the mind of the individual. He finds himself in presence of a great stock of acquisitions, which has been accumulated slowly by the community into which he is born. (It must be remembered, of course, that we are occupied with Western civilization.) At first sight, the material part of this inheritance seems to the average modern quite indispensable; the conveniences of civilization, such as gas, well-paved streets, the morning paper, are so entwined with his life that on removing from them he feels as if a part of himself were missing. And emigration to the interior of Australia, or of the Canadian provinces, produces some effects not altogether looked for. The vanity of the settler puts these down to patriotism and love of his relations; it is rather the traversing of old habit that is to blame. He misses the various modifications which human thought and labour produces in its environment. It is not difficult to understand why towns take a long time to grow. The erection of dwellings for a whole population is completed but slowly, while its churches and public buildings are in civilized countries the fruits of long periods of artistic and constructive skill, and indeed carry their history in their faces. Railways, shipping, canals, factories, and machinery are also the obvious product of human labour. The facts in the case of the country, if less obvious, are even more impressive. The cutting down of trees and brushwood, the drainage of the fields, the removal of stones from some soils, and the manuring of others, the making of roads, require the continual labours of several generations. But sometimes human industry takes a wider sweep, and aims at nothing less than a complete transformation of its environment, as in the planting of trees by which the climate of whole districts is modified, the conversion of marshes into habitable country, as in the drainage of the Fens, the emptying of the mountain tarns in the neighbourhood of Rome, the Dutch embankments against the encroachments of the sea.

These various contrivances are in perpetual need of repair or renewal. In this process they are usually modified and sometimes improved. The skill with which this is done depends on the state of scientific knowledge and of mechanical ingenuity. Thus the ground of all this adaptation of the environment to human needs lies very largely in certain intellectual states of mind.

Society having thus arranged the stage upon which the individual is to act, also prescribes with more or less rigour the manner in which he is to play his part. It imposes upon him certain injunctions which determine the limits within which free action is permitted to him, and visits him with various punishments if he oversteps them. The free sphere

CONDITIONS OF MENTAL DEVELOPMENT. 53

thus left to him is still further encroached upon by the dictates of public opinion which steps in to supplement, and sometimes, though less often, to suggest law-breaking. Public and social injunctions, at first imposed with some show of reason, from time to time become unfitted to regulate the citizen's life. The emotional force which made their exercise easy to begin with, gradually becomes spent, and in consequence there is a depreciation of social customs almost comparable to that of material conditions of civilized life. The need of intellectual operations there gives way here to the need of moral ones.

Let us review very briefly that store of intellectual and moral ideas from which the actual world of social life is from time to time renewed. All contributions to this stock are made in obedience to some practical motive. Science has gradually reached its present development in close connection with the arts of life; geometry, as its name denotes, was connected with land-surveying, chemistry with the desire to transmute the baser metals into gold, and so on. One of the most fruitful sources of improvement in chemistry and metallurgy has been the art of war, and physics has advanced in compliance with the demands of construction. The arts of pleasure have given birth to other forms of intellectual activity; history took its rise in the simple lay sung by a minstrel at the chieftain's table. The conversation that whiled away the leisure of the Greeks was the source of the metaphysics and logic of the present. In each age a few leisured individuals have surrendered themselves to speculation, and have furnished a basis for the intellectual culture of the age which succeeded them, and so ultimately for the complex conditions of modern civilization.

But the very progress that is made from time to time, whether in moral ideas, science, or art, brings about a new state of things in which the solutions of the problems of the period just past are no longer adequate to the needs of the present. Hence it is not to be expected that the mental equipment of any epoch has more than a transient and relative value. What is of value is the opportunity for the intellectual and moral activity by which each age may settle its own questions. Though the tendencies of speculation trend in great sweeps towards some determinate future, we cannot prophesy from viewing their past course the precise form in which they will issue. Hence the utmost freedom must be permitted to individual dispositions. The influence of a conformity, whether in politics or religion, which is brought about by an external compulsion, produces disastrous results of which the Laws of Manu and the Hindu character are not the only example; and it is noteworthy that those who by a great effort throw off these and similar bonds, are deflected further from a middle course than others on whom a lighter compulsion has been laid; the rebound is proportionate to the previous constraint. When, however, the development of the individual has been left as far as possible to take its natural course, the moment at which he begins to think for himself is not necessarily signalized by a great divergence from the path which up to that time he had followed in obedience to the usual educational influences. Many of the habits of thought which at first sight seemed unreasonable to him, are found to be grounded in facts of practice which he had failed to see, or seeing, had failed to recognize their value. Those legacies of religious and political dogma in which the consciousness of the past expressed itself, may indeed seem an incumbrance to the moral and political reformer in his hastier moments; but they constitute an accumulation of the same kind as the rest of the social inheritance, and the scrutiny to which they are subjected, if not less searching, should at the same time be not less respectful.

20*. **Organization of Experience.**—It must not be supposed that the individual mind is merely a receptacle of external influences. As these enter into the mental life, they tend to be organized into harmonious systems of thoughts, feelings, and desires. We must mark off therefore the impression, or its image, from the idea of the same as taken up into systematic thought; the simple feeling from the complex sentiment into which it passes; the isolated impulse from the universe of desires (cf. p. 147) into which it may be admitted. In these processes, the individual nature, the germs of which have been described (§§ 17, 18), both finds a characteristic expression, and is itself developed further. Perhaps the most luminous illustration of this fact is furnished by physiological nutrition. The raw materials of experience are, so to speak, digested and assimilated, until they become an integral part of the mind's equipment. In the next chapter we review the modes in which experience presents itself before this process takes place. The chapter on *The Laws of Mind* shows generally how the crude impression is transformed into part of an organized experience. The remainder of the book follows out this process in detail.

CHAPTER III.

SENSATION.

21. Characteristics of Sensation.—The sensory organs will be considered in this chapter only so far as a reference to their structure is required, in order that we may understand the sensations which we receive from them. Generally speaking, they consist of arrangements by which external stimuli may be transformed so as to affect the sensory nerves in a particular manner (§ 10). Turning now from the nervous structures to the sensations themselves, there are four aspects under which the various classes of sensations may be regarded. First, their *quality*, by which we may mark them off from other classes; a colour is obviously different in kind from a smell. Secondly, their *intensity;* sounds may be either loud or soft. Thirdly, in the cases of some senses, those of touch and vision for example, we can distinguish sensations according to the part of the sense-organ which is exercised. This property of being different according to the portion of the sense-organ which is excited, is called *local character*. Thus a small patch of scarlet, which occupies one part of the field of vision, may be distinguished from a patch of similar hue which occupies another part. Fourthly, different classes of sensation possess a differing interest for us apart from association.

SENSATION.

People who are fond of music, doubtless, find more pleasure than others in musical sounds. Painters of some schools, again, manifest a special delight in colour. This difference of intrinsic interest is sometimes called *tone of feeling*. We shall see in the sequel how this last characteristic determines the nature of our emotions (§ 98).

By appropriate movements, the sense-organs can be brought into different relations to the object to which we are attending; the hand, the eye, the nose, the ear, may be set to work on this side or that, as circumstances suggest. This possibility of moving the sense-organs greatly increases their scope. Borrowing a convenient phrase of Prof. Croom Robertson's, we may denote this as the *active* exercise of the senses.

22. **Measurement of Sensation.**—In considering the several senses, we shall have occasion to remark upon their comparative delicacy, that is to say, the minuteness of the stimuli, and of the changes in them, of which they can take account. This is called their Discrimination. Thus by the ear we can discriminate between notes, which differ by less than a vibration per second. With the sense of smell we can detect the presence of substances in such small quantities that they elude ordinary chemical re-agents. The discrimination of a sense can be measured, then, both by the minuteness of the least stimulus to which it can respond, and by the minuteness of the least change in the stimulus which is perceived as a change at all.

The question has been asked: How far do differences in the intensity of our sensations answer to differences in the strength of the stimuli which cause them? This inquiry has been responded to as follows: In order that the intensity of a sensation may be increased in arithmetical progression, the strength of the stimulus which causes it, must be increased in geometrical progression. The law was first stated by the German, Weber, and it is called after his name. It is found to be only approximately true, and to hold good most closely for stimuli of a moderate strength. Among the several senses, that of hearing furnishes the best examples which confirm it. For instance, if a sound of a given degree of loudness produce a sensation of a certain intensity, then in order that a sensation may be produced of twice that intensity, the stimulus must be increased to four times its former strength.

If we produce a moderately strong stimulus, say sound vibrations, by an instrument which we can regulate, and then gradually increase the force with which those vibrations are produced, the resulting impression, in this case an auditory one, will become more intense. Sooner or later, however, a point will be reached at which any increase in the intensity of the stimulus will no longer be followed by an increase in the intensity of the impression. This upper limit is called the Height for the sense. On the other hand, the intensity of the stimulus may be gradually decreased below the point from which we started, and the answering impression will grow fainter and fainter, until the stimulus is no longer perceptible at all. This lower limit is called the Threshold. The threshold marks the minimum intensity of sensation, the height, the maximum intensity of sensation, of which a given sense is capable, and the difference between them measures the scope of the given class of sensations. It is found that the delicacy of a sense is proportionate to the lowness of the threshold; in other words, a sense that can begin to take account of very minute stimuli can also take account of very slight changes

SENSATION.

in their strength. On the other hand, a sense which is too dull to register stimuli of less than a certain strength, will also fail to register changes of less than a corresponding amount. The two kinds of discrimination thus vary together. When local anæsthetics are applied to the skin, the decreased sensibility shows itself both in the way in which stimuli must be increased before they can be perceived at all, and in the lessened power of detecting changes in their intensity.

23. **Common Sensations.**—If we look at a diagram of the cerebro-spinal axis, that is to say, the brain and spinal cord together with the nerves attached to them, we shall see that the special senses of sight, hearing, smell, and taste, employ but a small number of the sensory nerves (§ 11). The remainder bring impressions of touch, and also give rise to a large class of sensations which are sometimes called Common or General sensations. These are distinguished from the special sensations, in being caused by internal changes which do not inform us as to what is going on in surrounding objects. They "are very probably the result of affections of the afferent nerves in general brought about by the state of the blood, or that of the tissues in which they are distributed."[1] One important class is connected with breathing, and continually prompt the acts of inspiration and expiration; if the breath is held for a time, these feelings soon become prominent. Another class is connected with the digestion, and take the form of feelings, now of hunger, now of repletion. A third large class is connected with the muscular tissues; according to their varying states, we have feelings of muscular freshness of fatigue.

These common sensations form, so to say, the background

[1] Huxley, *Physiology*, p. 203.

of consciousness; they are the basis of the states of comfort or discomfort which make life pleasant or a burden. An injury to some part of the body, such as a bruised limb or stomachic complaint, casts a shadow in which objects presented to the intellect are involved. This whole class of feelings is vague and indeterminate in character. They pass slowly and in long rhythms from one pole of well-being to the other. The great waves now set us at a high pitch of elation, now at a corresponding depression. These rhythms often take weeks to accomplish, and are themselves varied by minor rhythms. The times of growing depression and of growing elation are checked by temporary reactions.

24. **Pleasure and Pain.**—Reference has already been made to the fact that the common sensations tend to pass into pain. The same holds good of the special sensations. As sensations approach the Height, they begin to have this quality, and if the stimulus goes on acting after the Height is reached, the pain-quality tends to overcome the special character of the sensation. In this case we suppose stimuli of the usual kind, but of excessive strength, to act on the nerve-endings, the nerves themselves being unaffected. The case is different when the nerves are themselves cut or otherwise injured, or when the surrounding tissue is inflamed or otherwise diseased; the pains caused by cuts, burns, bruises, inflammation, are of this class. In this case the structure of the nerve is temporarily deranged or destroyed. The various classes of sensations, general as well as special, as they pass into pain-sensations, begin to resemble one another much more than under ordinary conditions. For the functions are interrupted, by which excitations passing along different kinds of nerves are distinguished.

The species of pain may be classified thus: pains of

SENSATION.

craving, pains of fatigue, pains due to structural hurt. This division answers to the insufficient or excessive stimulation of nerve-structures and to their positive injury. Pains of craving include the restlessness which comes of insufficient muscular exercise. Where constraint is applied the desire to move may reach an uncontrollable agony. Those pains which indicate the need for breathing or food have already been referred to. Pains of fatigue are most familiarly illustrated by those of the muscles. Nervous fatigue may be due to the over-stimulation of the peripheral or central parts of the nervous system. As the fatigue is increased, the dull headache rises in intensity and begins to throb, the wearied eyes first become jaded and then begin to smart. There is further a feeling of lassitude which accompanies general exhaustion or weakness. Pains due to physical injury are of several kinds. They may be dull, throbbing, burning, smarting, sharp like a knife; or a large amount of pain may be crowded into a single moment of agony. Pains of this kind seem to arise in a different kind of nerve from the ordinary sensations. It is well known that when the skin or other sensible portions of the body are wounded, only a touch impression is received at first, and then the pain sensation follows and gradually extends itself.[1] This tendency of pain to seem to spread itself, is one of its best marked characteristics, and is probably due to the fact that the disturbance on reaching the brain is communicated to the neighbouring centres.

Within the limits of excessive and insufficient stimulus, the exercise of the physiological functions, and especially the reception of impressions and the exercise of the muscles, is productive of pleasure. In this theory of the threshold

[1] *Wundt*, vol. i. p. 410.

and the height, we have the modern counterpart to the theory of the mean, though indeed Aristotle does not apply it to the explanation of pleasure.

A derivative kind of pleasure is experienced when there is a transition from a more painful to a less painful state. These pleasures of relief are almost as keen as any. It is questionable, however, whether the opposite transition from a more, to a less, pleasant state is a source of pain.

The pleasure which accompanies the putting forth of energies, which have been repressed for some time, is peculiarly exhilarating, as is also that which attends the gradual restoration of the muscles after exhaustion, namely, the pleasure of rest.

The fact that pleasure characterizes the normal and unimpeded exercise of physiological functions of all kinds, implies that where it attends the special senses, it will be coloured by their special nature. That is to say, while all pleasures have something in common, they do not, like pains, overcloud the mental activities to which they are attached. There are, therefore, as many elementary pleasures as there are normal sensations. Whereas, pains in becoming more strong tend to merge into one or two kinds, whatever be the sense-organ affected.

25. **Muscular Sense: Sense of Movement and of Resistance.**—One class of the common sensations deserves to be singled out for their greater distinctness and their importance as sources of knowledge respecting the outer world; those sensations, namely, which accompany movement of the parts of the body, and pressure against some resisting object. These so-called muscular feelings are rather difficult to localize. For all that, they are delicate, and enable us to discriminate between small differences of weight, and

very slight differences of movement. The movements of the head right and left, and up and down, of the arms and hands in almost every conceivable direction, of the trunk round its axis, and of the legs pendulum-fashion, all these differing motions producing different postures of the body, and in the case of the legs locomotion—are accompanied by differing sensations of movement. The movement of the right arm, for instance, up and outward, has its own special concurring sensation which is associated with just that movement. The muscles, however, which move the eyes, are capable of the most delicate adjustments of all, and their movements are discriminated most finely.

We may become conscious of a movement of the optic axis through very little more than one minute (the twenty-one thousand six hundredth part of a complete revolution). Not only the extent but the duration and the rate of movement of the ocular muscles, and those of the hands and arms, are also capable of very subtle discrimination.

The sensations of resistance are obtained when we press any part of the body against some solid substance, or try to pull it towards us. Like the preceding class, these sensations are capable of being distinguished very minutely. Thus a difference of one twenty-fifth can be detected in comparing weights of about a pound.

Concepts of force in all its manifestations derive their meaning in great part from these sensations of resistance. In a like manner, concepts of distance are based on the combination of sensations of movement.

26. **Touch.**—The organs of touch are distributed over all the free parts of the body, and over the walls of the mouth and nose. By these are received the sensations of touch proper, namely, of pressure; the sensations of warmth and

cold are probably dependent on a separate set of sense-organs.

The sense of temperature is not equal all over the body; its intensity seems to depend on the thickness of the epidermis. It is very delicate on the cheek and elbow. The susceptibility of the skin of the finger to small changes in temperature, approaches that of a fine mercury thermometer. The proper temperature of the skin seems to form a starting-point, temperatures below which are felt as cold, while those above it are felt as hot. The skin can adapt itself to temperatures which fall above or below this level. Thus, if one hand be placed in a basin of hot water, and the other in a basin of cold water, and if both hands are then plunged into a basin of tepid water, this will seem warm to one hand and cold to the other.

The sensations of temperature which are received by way of the whole surface of the skin only enter into the central region of consciousness when they vary considerably above or below the mean intensity. Although they generally pass unnoticed, sensations of temperature none the less form a massive and important group of elements in all ordinary states of consciousness.

Touch sensations proper—those, namely, of pressure—are often combined with those of resistance. They can be separately observed in the sensations received from the trunk or limbs when resting on some solid surface. If the hand is laid flat on the ground, and a weight is placed upon it, sensations of pressure alone are received from the skin. The discrimination of the sense of pressure has been measured for various parts of the body. Weber found that by placing different weights on the two hands simultaneously, a difference of about one-third could be detected; whereas, if the weights were placed successively on the same hand, the least perceptible difference was diminished to about one-fifteenth. If a weight is lifted, the sensations of pressure which are still received from the skin at the point of contact, are supplemented by the sensations of strain received from the arm. Care must be taken not to confuse sensations of temperature with those of touch proper. When two portions of the skin are in contact, those sensations which are not due to the pressure of the two parts on one another will be referable to sensations of temperature.

Sensations of pressure, like those of temperature, form an important element in consciousness. They are constantly being received, whether we stand, sit down, or recline, from those parts of the body on which we are supported. When the skin of the foot is benumbed, and the guiding sensations of pressure are no longer felt, the patient becomes giddy and totters.

In combination with the movement of the skin, sensations of pressure take a number of special forms. The following apparently simple sensations are due to sensations of movement and resistance combined with those of pressure. The list is quoted from Mr. Spencer's *Principles of Psychology*, Pt. VI. chap. xii. "The opposition which objects offer to compression or tension is distinguishable not only in its relative amounts, heavy and light, but in its kinds, hard and soft, firm and fluid, viscid and friable, tough and brittle, rigid and flexible, fissile and infissile, ductile and inductile, retractile and irretractile, compressible and incompressible, resilient and irresilient, and, combined with figure, the rough and smooth."

We have seen how muscular sensations can be distinguished according to the part of the body from which they are received (§ 25). The touch sensations exhibit this local character in a still more developed state. Sensations received from neighbouring parts of the skin are more easily confused together than those received from points separated by a greater distance. The distance at which impressions received on neigbouring parts of the skin are no longer distinguishable, and therefore appear as one, differs on different parts of the surface. This fact may be made the subject of a simple experiment. If the points of a pair of compasses are slightly blunted, and are then placed on the skin, the distance they must be apart, in order to be felt as separate, is least where the discrimination of the skin is keenest, as on the tip of the tongue, the lips, and the tips of the fingers. The distance by which the points must be separated, in order still to give rise to a double impression, rapidly increases as we pass to the less discriminative parts of the surface, until on the back, they must be at an interval of more than one inch.

27. **Taste and Smell.**—Leaving those sense-organs, which are distributed over the immensely greater part of the surface of the body, we now turn to the highly specialized organs placed in the neighbourhood of the cerebral hemispheres; those, namely, of taste, smell, hearing, sight. And first as to taste and smell.

Since they have to do with the chemical qualities of things, these have been called the chemical senses. They stand, therefore, like sentinels to challenge the substances which enter the stomach and lungs. Tastes always, and smells often, combine with tactile sensations of the tongue and mucous membrane of the nose respectively. Taste depends on a mechanical condition; substances, in order to be tasted, must be dissolved. Tastes have been classified into sour, sweet, bitter and salt. By combination of these simple flavours, with certain textures or grains, astringent, fiery, and other flavours are produced. Allowance must also be made for the concurrent effects of odours. "When the sense of smell is interfered with, as when the nose is held tightly pinched, it is very difficult to distinguish the taste of various objects. An onion, for instance, the eyes being shut, may then easily be confounded with an apple" (Huxley, p. 211). On the other hand, the potato is said, when sound, to have no proper flavour, and to owe its character to its texture (mealiness). By means of the taste very dilute solutions of sugar, salt, &c., can be detected. It is said that a sugar solution must be stronger than a salt one in order to be tasted, and that bitter or sour ones may be more dilute than either (Wundt, i. 372). Although taste has little of that definite character possessed by touch, sight, and hearing, it can be educated up to a certain pitch, as the examples of the cook and wine-taster show. It is the basis of the keen, though transitory, pleasures of the table, and often plays a large part in the interest of life, at a time when other pleasures are beginning to be less relished.

Smell is associated less than taste with sensations derived elsewhere; like taste, it only reacts on substances in a certain state. Substances, in order to be smelt, must be inhaled; they must, therefore, be in a gaseous form. If the breath is held, all smell ceases. This sense is exceedingly acute, and can detect quantities smaller than can be detected by ordinary chemical re-agents. Thus, a small particle of musk will scent clothing for years. Smell usually gives notice of the presence of decaying and other noxious matter. It also indicates the presence of certain kinds of vegetation and of some animals. Odours often gain a derivative interest by being associated with impressive occurrences, a certain persistence apparently fitting them for this office of symbolizing events with which they are associated.

28. **Sight: Binocular Vision.**—The organs of vision deserve a more detailed description than the other sense-organs. Their functions are better understood, and a slight knowledge of their structure will explain many points in vision.

The eyeball is not perfectly spherical; the lens forms a protuberance in front. The retina is the interior covering of the back of the eyeball; it consists of the continuation and extension of the optic nerve which, like the stalk of an apple, penetrates the eyeball from the back, pierces its outer coats, and spreads out on all sides, so as to form a kind of nerve carpet. Rays of light falling on this surface stimulate the ends of the nerves, and the nervous processes so set up give rise to sensations of light.

There are two points in the retina which deserve a passing notice, the yellow spot and the blind spot. The blind spot is at the entrance of the optic nerve. That it is indeed blind to light impressions may be learnt by a simple experiment. Put a dot and a cross upon a sheet of white paper about four inches apart. Close one eye, and fixing the other upon the cross, move the paper slowly backwards and forwards about a foot from the eye. At some stage in this process, the dot will disappear for a moment from the field of vision. The explanation is that it has passed over the blind spot.

The yellow spot is a slight depression in the centre of the retina, and is the part by which we see most distinctly. If the gaze is directed to an object, the eyes are moved until the image of it falls on this spot.

The lens resembles a very convex burning-glass, but is more convex on the inside than the outside. In the living eye it is clear as crystal, and consists of a somewhat soft substance, which becomes harder towards the back of the lens. The small chamber in front of the lens and the cavity of the ball are filled with transparent "humours." The rays of light have thus to pass through these humours and the lens before they reach the retina. Under ordinary circumstances they are so refracted as to unite into a distinct picture upon the background of the eye. The images so formed are inverted; the top of a tree is seen by the lower part of the retina, its base by the higher. Things to the right are imaged on the left, and *vice versâ*. A simple experiment will prove this. Press the outer surface of the eyeball where it projects from the orbit, and a figure like the "eye" on a peacock's feather will be seen, as we think, on the inside of the eye, but in reality on the outside, that being the place where the pressure was produced.

Now for another experiment. Will the reader be good enough to hold the forefinger of one hand near his right eye, having previously closed the other, and then put the forefinger of the other hand about fifteen inches further away. If the gaze is fixed on the near forefinger, the further one will seem indistinct, and if the gaze is fixed on the further one the near finger will appear indistinct. This experiment shows that the eye can be adjusted to receive distinct images from objects at varying distances. Now the rays of light falling on a convex lens from more distant objects, come to a focus sooner than the rays of light from nearer objects. Thus if the retina is to receive the distinct images of objects at varying distances, there must be a contrivance by which the focuses of the rays from those objects may all fall on the retina. And the manner in which this is brought about, is by altering the thickness of the lens. When it becomes thicker, the rays converge sooner, and the images of nearer objects no longer have their focus behind the retina, but upon it. Similarly, when the lens is extended, and thereby made thinner, the rays from more distant objects which, did it remain the same, would come to a focus in front of the retina, have their focus removed further away by the thinning of the lens. Thus in each case a distinct image is formed upon the retina; the increased distance with its nearer focus being compensated by the flattened lens and the consequently further focus; the lessened distance and the further focus being compensated by the thickened lens and its consequently nearer focus.

The light which we perceive is not all of one kind, and its various qualities constitute the different colours. The differences in quality are connected with the varying rapidity of the light undulations. These increase from the red end of the spectrum (400 billions per second), to the violet end (760 billions). The rays beyond these limits do not affect the retina. Thus the gamut of colour is barely an octave.

The colours of the spectrum are violet, indigo, blue, green, yellow, orange, red. By the mixture of these in varying proportions the different secondary colours, purples, browns, grays, are obtained. The mixture of coloured rays produces different combinations from that of coloured pigments. When, for example, a blue powder is mixed with a yellow one the product is a green powder. It used to be thought, therefore, that yellow light mixed with blue would give green. This is not so; when yellow and blue rays are blended, they produce a light that is almost white. The colour of an object is the result of a process of absorption; some colours, out of the light falling upon it are absorbed, and the rest of the light is reflected. Thus, blue objects absorb all colours of the spectrum except blue, yellow objects absorb all colours except yellow.

It has been attempted to derive all colour sensations from the composition of a few primary colour sensations. The theory which is known by the name of Young-Helmholz, refers all colours to the excitations, singly or together, of three kinds of retinal elements, answering to red, green, and violet rays respectively. This explanation of the colour sensations has been suggested in great part by the phenomenon of colour blindness. About one in twenty-five of the population is affected by this defect. It falls into two chief kinds. In one case, red and green are confused with one another and with gray, while the colours at the violet end of the spectrum are marked off in the usual way. In the other class, which is much rarer, blue and yellow are not distinguished as different colours, but only by their differing luminosity. Facts like these point to a distribution of function among different nerve elements of the kind suggested. And the different kinds of colour-blindness would thus be referred to the absence or imperfection of one or other of these elements.

In addition to differences in colour, visual sensations are marked by local character. The impressions made by a luminous point as it passes in front of the eye, are different for each point in the retina which it excites. This may be compared with the difference we perceive in impressions received from different parts of the skin. Further, we can distinguish sensations received by one eye from those received by the other.

Turning now from the quality to the amount and intensity of visual sensations, the nature of the retina makes it possible to distinguish sensations according to the extent of the retina excited. Larger objects cover larger areas, smaller objects smaller areas. The grouping of the elements simultaneously excited constitutes the form of the retinal image of an object. The keenness of retinal discrimination may be measured, first, by the smallness of the object which can be detected; secondly, by the least variation in its size that can be noticed.

The intensity of visual sensations depends on the intensity of the illumination. This may vary from the brilliancy of the sun's disk to absolute gloom. But the intensity of the sensation does not keep pace with that of the stimulus. The sun's disk seems equally illuminated to the naked eye, although the centre is forty times as bright as the circumference (according to Arago), and much more according to other astronomers. Thus the Height[1] for vision is far below the brightness of the sun's surface. Owing to the fact that the retina is excited even in complete darkness, and that we thus receive subjective impressions at all times, it is almost impossible to determine the threshold for vision, that is to say, that point at which illumination just becomes perceptible. The retina adapts itself to varying degrees of illumination, as can be observed by suddenly passing from the sunlight into a dark room. Thus the Height and the Threshold are relative terms;

[1] *I.e.* the most intense impression of light of which we are capable (*see* p. 58).

they differ with the brilliancy of surrounding objects and with the consequent retinal adjustment. The least perceptible change in illumination has been found to amount to about one hundred and twentieth part; that is to say, an increase or diminution in that proportion in the brightness of an object can just be noticed.

Thus side by side with the scale of colour there is one of brilliancy, and we can distinguish colours of the same kind according to the degree of their luminosity. There is also a difference in brilliancy among the colours of the spectrum themselves. Yellow, for instance, is more luminous than blue. Different tones of the same colour are produced by mixture with white light. When a colour is quite free from the admixture of white light it is said to be saturated or pure.

The sensation caused by a luminous appearance does not cease immediately the appearance is removed; it persists for about one-eighth of a second. This explains why falling meteors seem like strings of flame. When, however, the stimulus is strong enough to weary the retina, it leaves an after-effect which takes the form of positive or negative images. When we have looked at a bright object, a flame for instance, for some time, on turning the eyes away a bright image seems still to float before them. The excitement of the nerves continues after the stimulus has ceased, and gives rise to a *positive image*. If the stimulus, however, has acted for a considerable length of time, the retina becomes fatigued, and can no longer respond to it; for looking at a very bright light gradually renders the retina insensible. Thus on turning the eyes away, an ordinary degree of light fails to excite the over-excited portions, and the portions of the visual field answering to them appear as dark patches in a light ground; these are called *negative images.* If we have been looking through the window-panes at the sky on a very bright day, the window-panes still show as dark patches when the eyes are averted, the corresponding parts of the retina being fatigued; while the sash bars having protected the intervening parts of the retina, the before mentioned dark patches seem to have bars of light between them.

During waking hours visual sensations furnish a stimulant to nervous activity which is none the less important for being so gentle and uninterrupted. Any seclusion, even temporary, from the exciting effect of light-rays, lowers the nervous tone. Visual sensations have also a particular connection with the balance of the body. We are supported by the visible presence of solid objects in our immediate neighbourhood, and when these are missed, as in walking along the edge of a cliff, or on the top of a high building, our balance at once becomes less sure.

29. **Hearing.**—The sense of hearing, like the sense of sight, has a double organ, and we can distinguish between the impressions received by way of the two ears, and this fact gives hearing to a local character somewhat like that of touch and sight. Sounds are heard most distinctly when the opening of the ear is turned in their direction; this is done in listening, much as the eye is turned until the image falls upon the yellow spot. The external ear slightly aids the process of hearing by concentrating the sound.

The vibrations of some elastic body, usually the air, are the stimuli which give rise to auditory sensations. According to the regular or irregular character of those vibrations, auditory impressions take the quality of musical sounds on the one hand, and noises on the other.

The difference in the quality of musical sounds, or as it is sometimes called, the pitch, corresponds very accurately to the difference in the rapidity of the vibrations which give rise to them. The lowest tones are caused by very slow vibrations, about twenty per second. As these increase in rapidity, the pitch of the sound produced rises. The highest musical sound which can be perceived as such, answers to about forty thousand vibrations per second. The so-called musical intervals correspond to definite numerical relations between the rates of the vibrations which correspond to the notes forming the intervals. Thus the octave above any note answers to twice as many vibrations, the fifth to one and a half times as many, and so on. The middle C of the piano is produced by 264 vibrations, the G above it by 396, the next C by 528, while the C of the bass clef answers to 132.

It is a familiar fact that the tones of different musical instruments and of the human voice are distinguished from each other by their quality. If, for instance, the middle C of the piano is sung, and also struck or sounded upon the piano, violin, flute, trumpet, or other instrument, a note will be obtained on each occasion making 264 vibrations per second, and yet the various results are perceptibly different. The name timbre is given to this difference. It is due to the presence of overtones; a trained ear, for example, can detect in the middle C of the piano the note one octave above. And it is surmised that almost every musical instrument has its fundamental tones more or less accompanied by over-tones or harmonics. The number and intensity of these varies with each instrument, and gives its characteristic quality. Tones which seem hollow or empty, like those of a tuning-fork, are such as have few or no harmonics. A distinction has therefore been made between simple and compound notes. A simple note consists of one fundamental set of vibrations unaccompanied by harmonics. A compound note has both a fundamental tone and harmonics as well.

Concords or discords are produced according as the vibrations of two notes forming a musical interval do or do not coincide. Thus, in any octave, each vibration of the lower note answers to every other of the upper note. Where, however, the vibrations of the two notes coincide but rarely, as in the interval of the second (between two consecutive tones) the dissonance becomes marked.

The intensity of a sound varies with the force which sets the sound-medium in vibration. The keenness of discrimination for varying intensities remains approximately constant for a large range. It is found that the ear begins to detect a difference in intensity, when the strength of a sound is diminished or increased one-third. Test experiments have been made with iron balls allowed to fall from various heights on a vibrating surface at fixed distances from the ear.[1]

In opposition to the regular vibrations which give rise to musical sounds, stand the irregular ones which produce the sensation of noise. Sometimes these approach a rhythmical character, as in the noise made by a great waterfall. At the other extreme is the babel of a fish-market with its confused dissonances. The varied character of noises depends partly upon the strength of the separate shocks, partly upon the rapidity with which they follow each other, partly upon the presence of musical tones of various pitch which may be intermingled with the noise. Low tones are often blended in a grating noise, high tones with a hissing noise. Noises are most especially dis-

[1] *Wundt*, vol. i. p. 364.

tinguished from musical sounds by their intermittent character; the effect caused by interrupted sounds is characteristic, as in the beating of a hammer on the anvil.

As we have seen in the case of the other senses, so in the case of the ear, the degree of stimulation to which we are ordinarily exposed enters into our conscious states as something habitual and therefore to be expected. The townsman with the hum of town-life ringing in his ears is almost oppressed by the quiet of the country, while the farmer who comes up to town is stunned with the never-ending roar.

30. **Motor Impulses as an Element in Consciousness.**— There is one very important element in consciousness which is sometimes overlooked, sometimes confused with other elements. Many motor impulses as they leave the brain, make marks in consciousness which should not be confounded with the subsequent muscular sensations. The muscular sensation implies the passage of the motor impulse to the muscles, and the return of the sensory impression to announce the execution of the motor impulse. Thus a complete nervous circuit from the brain to the muscles and back again, intervenes between the motor impulse and the muscular sensation. But the first is no less present to consciousness than the second. This out-going motor impulse coincides under ordinary circumstances with the setting free of nervous energy; at least that is the external and mechanical way of putting the matter. But from the inner side, this putting forth of energy is the result of *our* activity. Thus the account of the senses needs to be supplemented by a reference to motor impulses before it can be accepted as exhausting the elements which enter into a state of consciousness.

This motor side of consciousness, determined as it is from within outwards, and not as in the case of sensation from without inwards, helps to give their specific character to the concepts of will and personality.

31. **Impressions of relation.**—There is one great class of impressions which does not fall under the head of any of the simple senses, and yet has a simplicity of character which justifies its being mentioned along with them. I refer to the impressions which accompany the changes from one sensation, or group of sensations, to some fresh sensation or group of sensations. Every such change is felt as a shock of more or less intensity. Thus the transition from silence to a loud sound produces a shock in us, quite independently of our noticing either the silence or the sound. When the transition is from one sensation to another of equal intensity, the resulting impression is very slight. The intensity of the shock depends greatly on the extent to which the mind is ready (pre-adjusted) to receive the fresh sensation. When we are unconscious of any shock in passing from one sensation to another, we assert their likeness. The violinist tuning his instrument to accord with the piano, the dyer matching his colours, goes on until he can pass from note to note, from tint to tint, without a break in consciousness. On the other hand, if the shock is great, we say the impressions between which we pass are unlike, and the intensity of the shock is a measure of the difference. The shock of unlikeness is also felt when we pass from an idea of a sensation to an actual sensation, or from one idea to another idea. Readers of *Vanity Fair* will remember Miss Sharp's adventure with a chili. "How fresh and green they look," she said, and put one in her mouth. It was hotter than the curry!

CHAPTER IV.

LAWS OF MIND.

32. Constituent parts of a State of Mind.—The sensations treated of in the last chapter are ordinarily due to external causes, or states of body as we decided to call them (§ 8). We shall now attempt to trace the laws according to which states of mind give rise to other states of mind. We shall first observe the different elements which reveal themselves in a single state, and then follow them into the succeeding state, and take note of the different shapes assumed by them and the effects which they produce. If at this moment the reader withdraws his thoughts from the outer world and fixes them on their relations to each other, the following elements will reveal themselves to him with more or less force and in varying proportions:

I. (*a*) Sensations received by way of the special sense organs, such as the eyes and the skin.

(*b*) Sensations from those parts of the body which lie below the surface, and especially from the organs of respiration, circulation, and digestion: these are not very distinct, and form, as it were, the background for the rest of the sensations.

(*c*) Motor impulses tending to set this or that set of muscles in motion.

II. (*a*) Echoes of similar sensations received on previous occasions, that is remembered sensations. These may be grouped together in two ways. Their order may be the same as that of the first experience. We may be thinking of the events in a country walk in the order in which they were received. Sometimes, on the other hand, these memories are put together in other ways than those of the actual experience. We may associate the colour and lustre of gold not with the shapes in which it is most familiar, but with the form of an egg; readers of fairy tales go through some such process of thought as this, while they picture the narratives to themselves.

(*b*) Certain symbols, which are usually words spoken heard or read, set the mind to follow out certain trains of memories. The result of following out one of these trains of memory is a concept, and the whole process is called conception, or grasping together, the sign being, so to speak, the magnet which holds the associated memories together.

III. The same process of remembering, re-arranging, and combining memories of past operations can be traced out in the case of the feelings. Remembered pains and pleasures are grouped into complex trains of emotion, and we find side by side in the same state of consciousness simple feelings, and combinations of feelings in process of development from them.

IV. Lastly, trains of actions or tendencies to action, some of them realizing themselves at the moment, others the echoes of similar trains of action previously carried out, complete the tale.

We have thus traced out the main divisions under which we might range the elements of a state of mind. As we have already seen, all classifications must be rough; each train of operations implies the activity of nearly every side of the mind. Hence in attempting to make a classification, mental operations are grouped with reference to their most striking characteristic, and calling them by this or that name must not be taken to imply that they have not other characteristics. We call a train of thought intellectual because that is its most striking feature, but each train of thought is also emotional and involves the will.

Of course all these numerous elements could not be observed in any single act of inward gaze. We must therefore take the list as representing what we might detect if that inward gaze were ten times as keen as it is.

33. **Their Transformation into the Succeeding State.**—Now let us look inwards again: the elements which composed the state of mind just past are changed. Some have vanished, and are replaced by fresh ones; others have remained, but stand in an altered relation to the rest. If in the interval the eyes have been turned through an angle of forty-five degrees, part of the visual field will be visible no longer; its apparent place will have been taken by the pictures of objects in the adjacent part of the field, and these in turn will be followed by new pictures on the side vacated by them; the whole process resembles that by which a panorama is exhibited. Similar changes take place in every part of consciousness; alike in memory, conception, feeling, and will, old elements begin to disappear from the stage, and new ones to come on. Leaving the golden egg, we think of the goose that laid it; or the memories of the country walk may pass away to make room for memories of the friend with whom it was taken, and we may then think of him as pursuing his usual business. The process of conception just begun may now be practically

G

complete; the word statesman, for example, may have been followed out through all its associations, and we may have before us an abstract of the common qualities of our favourite party leaders. Or some connection between an old train of thought and an isolated circumstance may flash upon us; and we realize with a start how the mind may pass over the most obvious bearings of its thoughts one on another. This last-mentioned process may serve as an example of the way in which judgments are formed.

It is this passage from one thought to another which best exhibits the mind's true nature. Hence in the pages which follow we shall look at each thought as the cause and source of succeeding thoughts, no less than as interesting for what it is in itself. This transition from one state of mind to another gives rise to, and makes itself known by, those elementary impressions of likeness and unlikeness described in § 31. All the processes of comparison, whether of sensations, as illustrated in the last chapter, or of ideas as parts of intellectual processes, depend on the fact that we are conscious of the transition from one state of mind to another, no less than of the states of mind themselves.

34. **Law of Persistence: Habit.**—If we look at a piece of burning wood whirled rapidly round, the eye does not follow the flame, but receives an impression as of a ring of fire. The sensation caused by the light coming from the flame at one point in its path, does not cease, before it is followed up by similar sensations caused by the flame in the succeeding points of its path. In this way earlier impressions persist while later ones are being received. The same holds good of all operations of mind, and not merely of visual sensations. Each mental operation leaves a trace.

This trace, strong at first, gradually becomes weaker, until it is no longer perceptible. The visual impression remains vivid for a fraction of a second, and then becomes a faint memory and nothing more. The thoughts and feelings of a time long past usually leave effects so slight that we are scarcely affected by them. This diminishing power of impressions may be due to the cessation of the effect (sensory impression for example), when the cause (the sensory stimulus) is removed; in this way actual impressions become faint and pass away. Or a train of thought may lose its power and vividness through the predominance of a conflicting train of thought. That failure to repeat trains of thought is chiefly due to their being impeded by later acquisitions, is shown by certain affections of memory. When it begins to fail, the last acquisitions are the first to go, and in so doing they lay bare the earlier ones. The old man who has a worse memory than others for recent events, has for that very reason a better one for earlier events.

But the trace left by a mental operation is much more than the mere persistence of its effects. If this were all, there would be no reason to emphasize a quality which marks all facts. This persistence, in the case of mental operations, involves a disposition to go through the same operation again. To have had a series of thoughts, of feelings, of intentions at any time, means that for the future we shall more readily have the same thoughts, feelings, and intentions again; so that on the side of the intellect, this law explains memory, on the side of the emotions, temperament, on the side of the will, habit and character.

35. Mental Phenomena as Products; Processes; Tendencies, or Dispositions.—In the first paragraph of this chapter the elements of a mental state were regarded as they present themselves to our gaze. From this point of view they are called presentations. In the second paragraph, we saw how presentations are transformed into, or make way for, succeeding presentations; that is, we saw mental operations in process. (It is useful to distinguish between the processes and the products of thought; thus, percepts and concepts are the products of the processes of perception and conception.) In the last paragraph, we have seen how each mental operation by its mere occurrence constitutes a tendency or disposition to go through the same operation again. Here we have three aspects of the same facts, none of which is to be lost sight of. We may be helped to understand them better if we watch them varying in scope and intensity. In the first place, a state of consciousness viewed as a product may be very rich and full; such is the concept of English History to which a great historian attains after years of labour and reflection. Or a mental state viewed as a product may be poor and empty; our ideas of objects to which we have given little attention are such. (The amount of meaning contained in a mental state is called its *content.*) In the second place, a state of consciousness may be a stage in a process, and as such may be regular and complete, or irregular and incomplete. When the stages in a mental operation follow one another regularly, and when each element involved is brought to view, the resulting product is both full and stable. Opinions which rest on a wide and careful consideration of the facts are stable and comprehensive. Those, on the other hand, which have involved the rising to view of few presentations, are incomplete and unstable; the impressions of a place derived from one or two short visits are of this kind. In the third place, a state of consciousness regarded as the basis of a disposition may be strong or weak. Faint impressions leave no decisive mark upon our minds either as regards the present or the future.

Observe that these several characteristics do not necessarily rise and fall together. A state of consciousness may be full and faint, or it may contain few elements and yet be vivid. A man may be dominated by a single idea, and may give it vigorous expression; he may be filled with varied thoughts, and yet be unmoved by them. We can distinguish, therefore, the fulness of a thought from its activity; its presented content from the energy of the process of presentation, or to use a convenient phrase, its presentative activity. The first aspect is the intellectual side of a thought, the latter aspect the emotional side. Hence a mental process, of which intensity is the chief characteristic, may be classed as emotional; while one which unrolls itself before the mind without disturbing other elements to a serious extent, is intellectual. Since these other elements are not disturbed and thrust under, the mind can trace the relations of the new thought to those whose company it has joined, and so the mutual attractions and repulsions, the likenesses and dissimilarities, in which the special object of intellectual operations is found, come to view.

36. **Combination of Mental Elements: Association.**—When a series of mental operations has been performed, what remains in the mind is the disposition to repeat that series on occasion being given. The repetition of a train of mental operations, therefore, is not the act of reference to memory's strong box, which some popular metaphors suggest. Past experiences are not stored up like the files of old newspapers. We can only recall an experience in proportion as at the moment of recall we are capable of

undergoing the like experience again. That is to say, the revival of a series of mental operations depends on the elements present in consciousness and the way in which they stand to one another, at the particular moment of revival. At the beginning of this chapter, an attempt was made to enumerate the elements, which, in combination, constitute an average state of consciousness. It was pointed out how by one set of activities several attributes associated with the word gold were combined with those of an egg; how the various imagined qualities of the leaders of one's party were used again, but in a slightly modified arrangement, to form the concept of a statesman; how the memory of our friend now forms part of the recollected country walk, now part of another set of circumstances. Thus, the same elements are for ever being taken out of old combinations, and built up into new ones. The various combinations into which they enter may be reduced to their elements, much as at first the perceptions of surrounding objects were reduced to *their* elements (§ 1). In the mental life, therefore, perceptions images concepts feelings resolves come and go; the tendencies of which they are the expressions alone remain, together with the small outfit of a few elements, enumerated in § 32, in whose interactions at first, second, or third remove from actual experience they consist. That out of a collection thus limited, the infinite varieties of human thought should be built up, is not surprising when we take into account the various degrees of intensity exhibited by each of these ultimate elements, and remember how music, out of far fewer elements, obtains an infinity of different combinations.

Let us now see how, by the changing forms taken by these mental combinations, occasion is given from time to time for old tendencies to be followed out again. Each stage in a train of operations tends to call up that which on one or more previous occasions has followed it. Now, as we have seen, the same element, or combination of elements, may recur in different combinations. In this way each element, or combination of elements, presented to the mind tends to call up with varying degrees of force, all those other elements or combinations which have followed it on various occasions in the past. Thus, suppose that we have had a succession of experiences denoted by the letters a, b, c; then whenever a is presented, b will tend to follow it, only in turn to be replaced by c. Suppose that we have also undergone a series of experiences symbolized by the letters a, d, e; then whenever a is presented, d and e will tend to follow it. If now a is caused by whatever means, it will tend to be followed by b and c on the one hand, and d and e on the other. The patch of red down the street (a) may suggest the approach of a regiment (b), and its march past (c); or it may suggest the banners of the Salvation Army (d), and a religious meeting (e). These two alternatives rise to mind because the sight of a patch of red has been followed in the past by the experiences referred to.

When a series of impressions has been invariably received in one order, the revival will follow the same order, unless by a strong effort of the attention we check this tendency. The difficulty of inverting a familiar experience may be measured when we try to repeat the alphabet backwards. Other series are presented to us now in one way, now in another, as, for instance, the musical scale; these can be gone over either way with equal ease. Association also holds good between the elements which compose the same state of consciousness. If one circumstance is recalled, the memory of the accompanying circumstances comes into play. The reception of an idea is more or less associated with all the circumstances in which it was received, the speaker's bearing and accent, and our own place in the audience. There are three ways, then, in which association acts— forwards, forwards and backwards, and sideways or laterally, between elements of the same state of consciousness.

37. **External Conditions of Revival.**—It is but seldom that the mind is completely buried in its own workings, and that mental elements weave themselves into new combinations without external interference. The states of the external sense-organs, changing in obedience to external stimulus, and of the internal sense-organs influenced by physiological conditions, offer fresh points of departure from time to time to the processes of reproduction. Immediately one train of thought has been suggested from without, a second external suggestion disturbs it and introduces some other train. Practice in thought may avail somewhat in withstanding these external influences; it can never nullify them. Hence pure thought is an unattainable ideal. We must always allow, then, for the effect of this unavoidable bias, in estimating the accuracy of our conclusions. Indigestion and bad weather are psychological, if not logical, bases of pessimism; while your citizen, who has just risen from a good dinner, suffused with the consciousness of a large balance at the banker's, is inclined to under-estimate the privations of the unemployed artisan.

38. **How the Strength of a Tendency is Determined.**— The capacity of consciousness is limited; there is not scope in it for all the combinations of elements which tend to occur. The mind, at each instant, is dominated by one

or two series of processes, while other series fail to be enacted, and only one or two are faintly recognized by the side of the predominant ones. There is thus a kind of survival of the fittest, by which the weaker tendencies are prevented from coming to reality in consciousness. The force with which a tendency makes its way into consciousness, and resists expulsion, is determined by three things— first, its own strength; second, the strength of supporting tendencies; third, that of opposing tendencies.

I. The strength of a tendency by which a series of mental operations is gone through again, depends on the amount of the pleasure gained, or pain avoided, by its means. Interesting experiences are more easily recalled than others. It also depends on the depths to which previous experiences have impressed us at the time of their occurrence; striking events recur more easily to the mind than inconspicuous ones. It depends, too, on the number of times the operation has been performed before. Each process of intellect, feeling, or will, makes succeeding operations of the same kind more easy, until a certain stage is reached; at this point progress no longer follows upon exercise, and may even be thrown back. Reference was made to this fact in § 15. Thus a natural limit is set to the increasing strength of a tendency. Borrowing a term already used for sensation, we may call this the Height. This upper limit varies very much. Where the nature of the elements involved is intrinsically pleasurable, the limit to a disposition seems scarcely to exist, and the dispositions which repeated exercise forms, become so strong that they are persisted in when they no longer bring any pleasure at all. The liquor habit shows how acts at first pleasurable, are repeated when they no longer bring any pleasure. Below this limit the strength of a disposition to pursue a train of mental operations, depends on the number of times that train has been followed out in the past. It also depends on the way in which those occasions have been distributed in time. The repetition of an operation at long intervals is less effective than the same number of repetitions at shorter intervals. The strength of a tendency also depends on the recency of previous exercise; hence the training athletes undergo before competing in a race, or other contest of skill; hence, too, the effort made by a student just before an examination. The strength of a disposition also depends on the plasticity of the mind at the time when it is being formed. Those impressions are deepest which are received in the early years before the mind is seared and battered by experience.

II. We now consider the way in which one tendency may be supported by another. Those tendencies are strongest which are associated with pleasure gained or pain avoided; hence by association with them, other tendencies influence the mind, which when followed out are indifferent in themselves. In this way the actions, feelings, and thoughts, by whose operation some desired state of mind is attained, receive a reflected interest. Work of any kind is performed at first as a means to some end and for the sake of that end; later it may be found to be pleasurable in itself. At this stage the interest we take in it is doubled; it is due both to the intrinsic nature of the activity and to the end in view. The same train of thoughts, feelings, or actions, may lead to several desired issues, and from each of the latter may receive support. Or slightly changing the phrase, the same series of acts may be due to many co-operating motives.

III. A tendency may be neutralized by conflicting tendencies. Pain so fills consciousness that it excludes the tendencies which, in its absence, would realize themselves. Or again, the scope of consciousness is so limited, that the mere presence of some tendencies will prevent others from being developed. The state of mind called distraction is due to this inability to follow out more than a limited number of operations at once. In cases like these tendencies may exclude one another indifferently, and the opposition may be owing simply to their occurrence under conditions which do not permit them all to be realized. Some tendencies, however, excite opposing tendencies whereby they are neutralized. The dispositions to perform certain acts may be associated with the motives which would prevent them from being followed out. Most tendencies to self-control take this form.

39. **Combination of Tendencies in a Single Consciousness.**—These various series of operations, in which mental dispositions express themselves, have hitherto been considered separately. They do not, however, exist separately; they are combined into one process, namely, the life of consciousness, and form one experience. The fact of their being so bound together is referred to our personality. This does not help us very far; for if we are asked what we know about our personality, we can merely repeat the facts it seek to explain. This unity of the individual's consciousness exhibits itself in two directions. On the one side we find that the various constituent elements of consciousness at a given moment are connected together; on the other side, each state of consciousness is connected with that which precedes. Thus serial and simultaneous association may be regarded as different expressions of the

same thing. And the persistence of mental disposition, from which the laws of association are corollaries, is grounded on the same fact as they. The word *I* is the most familiar term by which it is denoted.

Making use of an analogy drawn from mechanics, the life of consciousness as a whole may be regarded as the resultant of these several tendencies. Further, there appears to be necessary a kind of balance between these contending elements, in order that the tendency of consciousness as a whole may not be too violently changed, or may not go too far in any single direction. Hence it must be able to answer a powerful impulse in one direction by a compensating impulse towards another. This self-compensation takes place in two directions. Ideas left to themselves combine into the most fantastic and unreal structures, and require the control of actual fact in the shape of sense impressions; dreams show us how far the mind can stray from a normal path when it is left without the stimulating guidance of the senses. At the same time, the immediate impressions and impulses of each moment are too crude and rough to suit the delicate requirements of our surroundings. They need to be filled out or corrected under the influence of past impressions and impulses.

40. **Predominating Tendencies: Attention.**—What happens when we attend to anything? Take the instance of a child learning a drill-exercise. If it is attentive, memories of play are driven out of its mind, and it is not attracted from the drill by what is taking place around it. Thus the act of attention to the process of imitating the drill-instructor's actions, implies that conflicting tendencies are driven from the mind, or prevented from entering it. Impulses go to the muscles which are to produce the

desired bodily movements, and contrary impulses are checked. Thus a process of concentration goes on, or in other words, certain tendencies maintain themselves to the exclusion of others. The same is true of more purely intellectual operations. The child, as it applies itself to a geography lesson, is occupied with images of sea, land, rivers and cities to the exclusion of other images. Attention, then, consists in the predominance of certain tendencies over others. By this predominance, they become more distinct and vivid. Let us examine the different forms which may be taken by this process. These differ in extent, intensity, concentration, and versatility. The *extent* of the process may be measured by the number of tendencies which may be distinctly and vividly present to the mind at the same time. The conductor of an orchestra is called upon to display this wideness of attention; string and wind instruments of various kinds have to be kept under his control. A second important aspect of attention is its *intensity*. This shows itself in the detection of minute details and in catching the relations of things one to another. A trained mechanic sees at once how the different parts of a machine stand to one another, while a person, without similar training, passes them over with scarcely a glance. The *concentration* of attention is shown by the control of conflicting tendencies. Even the distant cannonade of Jena could not withdraw Hegel from the composition on which he was engaged. By the *versatility* of the attention is meant the ease with which it can pass from one subject to another: this is the same as quick dissolution of old combinations and the speedy formation of new ones. It is manifested by practised journalists, who can treat in succession of the most diverse topics.

These four aspects are connected more or less closely with each other. The extent of the field of attention is opposed to its intensity. If it is directed to a large number of things simultaneously, no very precise idea will be obtained of any single one. Concentration implies the narrowing of the field of attention in order that it may be deepened. This fact is made use of by conjurers: they concentrate the attention of the onlookers on some point, and so may perform manipulations at other points without being detected, the spectators' attention being fixed elsewhere. Now concentration usually demands time for its completion, and cannot be attained at once. When the child is applying his mind wholly to a lesson in geography, it is filled with geographical ideas to the exclusion of others. All that it remembers of previous lessons rises to view, and supplies it with a fund of associations to which the teacher may appeal. It is clear that the coming into consciousness of so elaborate a group of elements is a lengthy process. The child's attention will not be completely active until some time after the class has begun. The contrary process of dissolving such a combination of ideas also demands time. Thus concentration is at first opposed to the faculty of passing rapidly from one train of thought to another. Only continued exercise can combine the power of deep concentration with that of rapid change of subject. In other words, versatility and concentration are often brought about at one another's expense. It is worth remark here that the attention tends to fix itself on one object at a time. Hence, when we seem to be attending to several objects at the same time, the mind is probably passing rapidly backwards and forwards from one to another. The extent of the attention has been made the subject of measurement. Thus it has been shown by experiment that not more than six different visual impressions, such as lines, letters, figures, can be attended to at once. If, however, these separate elements are grouped together in familiar combinations, many more can be made the objects of a single act of attention.

Mr. Galton tells an amusing tale which illustrates the limitations of a savage's mind. "When a Damara's mind is bent upon number, it is too much occupied to dwell upon quantity; thus a heifer is bought from a man for ten sticks of tobacco; his large hands being both spread out upon the ground, and a stick placed on each finger, he gathers up the tobacco; the size of the mass pleases him, and the bargain is struck. You then want to buy a second heifer; the same process is gone through, but half sticks instead of whole ones are put upon his fingers; the man is equally satisfied at the time, but occasionally finds it out and complains the next day."[1]

41. **Theory of Attention.**—The processes which at any one moment are going on in consciousness, are not all equally vivid and distinct. Some are very strong, others less strong, others barely perceptible. The activity of the attention is confined to those tendencies which predominate. Now these predominating tendencies are as much more vivid and distinct than the rest, as the central parts of the field of vision excel the outlying parts; attention is said, therefore, to work in the central portions of consciousness. But there are many very intense trains of feeling to which we can hardly be said to attend; since they go beyond a certain intensity, the balanced working of the mind is disturbed, and it surrenders itself to some particular

[1] *Travels in South Africa*, chap. v.

current of its experiences. In attention, on the contrary, the mind is occupied in tracing out how its various perceptions and thoughts stand one to another. When we attend to a group of such mental elements, they do not remain unchanged; since we are very vividly and clearly conscious of them (for this is what we mean by attention), the ways in which they stand to one another rise to view. As we con a problem in Euclid, the relations between the various parts of the figure become more obvious, and we become able to conceive the solution. Now this detection of the relations of our thoughts one to another, is only possible when the various processes can be brought together side by side for comparison. It is clear, then, that if any single process overpowers all the rest, we are incapacitated for observing how it stands to them. Attention may thus be described as being busied with those processes which are intense enough to be quite clear and distinct, and not so intense but that several of them may occupy consciousness at the same time. (*See* § 35, end.)

But the attention is more than the mere predominance of one or two trains of thought; it is a complex process, and implies several subordinate processes. Perhaps the most important of these is the control of opposing trains of thought, which would tend to interrupt the predominant trains. We have here in the central region of consciousness a kind of regulation like that which was mentioned in a previous paragraph, as holding good for consciousness as a whole (§ 39). The prime necessity is, that the mind shall never so be under a single influence but that it can release itself. This is the basis of all the theories of indifference to external motives, which have formed the staple of so many ethical systems. Attention is thus an *activity* of the mind, and manifests itself in two ways—positively in self-direction to some trains of thought, feeling, or action; negatively in self-withdrawal from others. The effort involved in attention makes a mark in consciousness like that described in § 30.

The attention also involves an emotional element. The name given to this is interest. The interest of a mental tendency is that which causes its power over us, and may therefore be measured by the strength of the tendency. Thus all those conditions which were enumerated in § 38 are really elements of interest. Interest therefore is merely another name for the extent to which a mental tendency makes itself predominant. This meaning is more obvious in the case of other words used to denote the same fact, such as attractiveness, impressiveness, influence. Association with pleasure or pain is an important element in determining the strength of a tendency, and therefore in determining its interest. Hence it is commonly taken as its whole meaning by a familiar looseness of thought, which mistakes the qualities of the chief species for those of the genus. But this is too narrow. Our idea of interest must take account of *all* mental conditions which help to give one train of thought or feeling the mastery among others. What are called the chief interests of a man consist in those associations and aims on which he dwells most frequently. It is a great mistake, however, to assume that he is merely governed by pleasure or by the desire to avoid pain in so doing. Habit is the great agent here, and all the conditions which give rise to and strengthen habit. How these combine in the production of emotional tendencies we shall see later on in the proper place. At the same time, association with pleasure or the avoidance of pain is one of the most efficient means of creating interest. Summing up, the emotional side of attention may be described as follows : mental processes must attain a certain intensity, and must therefore be marked by a certain degree of emotional colouring, in order that they may be attended to.

42. **Time taken in the process of Attention: Personal Error.**—The process of attention, by which a presentation passes from the vaguer to the more vivid regions of consciousness is not instantaneous. When an object, be it a sensory impression or an idea, has begun to affect the mind, it does not all at once force itself on the attention.

Things are sometimes quite ludicrously overlooked in spite of their being present to us ; the writer searches for the pen which he is holding between his lips, or the spectacles which he has pushed up on to his forehead. Thus it is quite possible to be receiving impressions without taking notice of them. The pen and the spectacles give rise to sensations of contact, but these fail to impress themselves on the attention. Gradually there dawns upon the mind the perception that the pen is between the lips, and the glasses on the forehead.

Like impressions from without, voluntary impulses pass through varying grades of vividness before they reach realization. First, there is the faint representation of the movements to be performed; this rises in clearness, and merges ultimately into the motor impulse on which the muscular contractions follow. All these processes of perception and action are performed at different rates by different persons; hence in comparing observations in which the time of an occurrence has to be registered, we must make allowance for such differences in speed. Thus in order that an astronomer may record the passage of a star across his telescope, the following five stages are required: (i.) the passage of the sense stimulus along the optic nerve to his brain; (ii.) the perception of that impression; (iii.) the connecting of the impression with the movements which are to record it; (iv.) the rising of the answering motor impulse into greater and greater clearness; (v.) the passage of this impulse to the muscles. The different times taken by different observers in executing this process give rise to what is known as the "personal error."[1]

The whole process is what has been termed a reflex (§ 13). It consists in the reaction upon a stimulus, and the time taken is called the reaction time. With repeated practice of any given reflex, it is performed more and more quickly, and involves the attention less and less until it is done automatically. The reactions have consequently been classed as complete and abbreviated, according as the attention is involved or not. In some experiments, in which the subject made signs in response to sound impulses, it appeared that a reaction of the simple kind described above took about one-fourth of a second in its longer form, and one-eighth in its shorter form. The difference between the two lengths of time is that taken by the attention. This amounts to one-eighth of a second. And, generally, we may take it that the self-direction of the mind to an external object, or an internal process, occupies about one-tenth of a second.[2]

43. **Attention and Expectation.**—Attention implies that the mind shall adjust itself to the impression received; we must have present to us the circumstances associated in our experience with the impression, or otherwise it will fail of its proper effect. Hence if we are buried in one train of

---
[1] Jevons, *Principles of Science*, chap. xv.
[2] Wundt, vol. ii. p. 316.

memories, sense impressions which appeal to another train, take some time before they can affect the mind. Witness the instances of the spectacles and the pen. It is this process of adjustment which makes us more slow in attending to an impression which is not expected. It takes about twice as long to execute a movement in response to a signal which is unexpected, as to respond to an expected one. Expectation may therefore be defined as preadjusted attention. The amount of this preadjustment may vary between two extremes. We may be in a state of mere expectation without any definite idea of the future event in our minds. From this state we may gradually rise to one in which not merely the nature, but the extent and circumstances, of the event are present to the mind. It is often amusing to forecast the surroundings of some one to whom we are paying a first visit, and to compare the reality with the idea. This state in which the idea of an event stands ready for realization, implies that those parts of the nervous system which the reality will affect, are in a state of excitement. Hence if we are expecting to have some mental or physiological experience, this mere state of expectation will sometimes realize itself in anticipation of the expected circumstances; the wish is often father to the thought. And even where this is not so, the strained attention deepens the effect of the impressions when actually received. Expected pains are far more keenly felt than those which take us unawares.

According to the extent of the preadjustment, the attitude of the mind is receptive or active. When we are awaiting an experience the nature of which is but little known to us, expectancy consists in the readiness of the mind to exclude everything that shall conflict with the due effect of any impression whatsoever; that is to say, we put away any fixed presuppositions as to the event expected. This should be the disposition of the child to whom his teacher is about to give a lesson on a new subject; it is also the state of mind in which scientific investigators should approach unfamiliar facts. Here the attention is *receptive*.

Active attention, on the other hand, implies that the mind has some idea of what it is expecting, and is therefore in a position to cast on one side indifferent circumstances. It is not merely ready to receive impressions : it possesses means to sift them. Not every impression will be remarked, but only those which relate to some particular subject. A large and more or less distinct group of memories rises to the mind, and we try to bring present impressions into connection with them; memories and present impressions correcting and supplementing one another. In listening to a lecture, those remarks make most impression upon us which contradict, or agree with, our previously formed idea of the subject; while those which have no analogy in our past experience, or do not stand in direct opposition to it, pass unnoticed. As we listen, the group of mental elements with which we started is developed and transformed under the stimulus of the successive details brought forward by the lecturer; each fresh remark, as it is heard by us, is brought into relation to this large group. But this process is always going on, and not merely in the lecture-room. Large and complex groups of associations, such as those connected with our home, business, and other familiar surroundings, are being transformed and enlarged constantly, and new ones also come into being. Active attention consists, then, in bringing new impressions into connection with these groups. And concrete terms like cloud, tree, flower, or abstract ones, like justice, patriotism, are symbols serving to set at work the tendencies by which the groups corresponding to those names come into the mind again. They compel attention, if our associations with them are rich and numerous. And so, in dealing with familiar things, each circumstance is rich with a force of suggestion by which the mind is made to gather itself together to tackle them; the mechanic working with familiar tools on materials the properties of which he knows, has countless memories of past experiences to guide him; he can forecast the result of each stroke of his tool; if he is a good workman, expectation is always passing into reality.

The process of adjustment is subject to two kinds of error. One is the tendency to respond before the actual stimulus has affected us; the other is the tendency to respond to the wrong stimulus. They are both due to excessive mental activity not kept under control. In the former case, the mere thought of the stimulus is enough to set the mind at work; the thought of a sound is often mistaken in this way for the actual sound. In the second case, any stimulus, even though it be other than *the* stimulus, is enough. Mrs. Wardle sen., in the *Pickwick Papers*, always referred startling impressions to the kitchen chimney being on fire.

44. The Effect of Novelty and Familiarity on the Attention.—When an impression has been received many times, it begins to lose the exciting influence which it possessed over us at first. This loss, however, is in many cases compensated by the associations which gather around it; familiarity does not always imply loss of interest (§ 41). The interest of an impression depends on the number of the ways in which it appeals to us. One, then, which touches us at many points will always be interesting, while one which only affects us by its novelty, necessarily loses its power when the novelty has passed away. The child longs for a new toy merely as offering a new impression, and casts it aside after a few hours' play.

45. Acts of the Attention as Measuring the Lapse of Time.—When life is spent amid surroundings which offer but little to arouse the attention, no deep marks are made upon the mind. And in the absence of such milestones, time, as we look back upon it, presents little to indicate the distance already traversed; the past, having impressed itself so faintly upon us, is forgotten even amid the uninteresting present. And yet, although the past slips away so easily, the present seems to linger; the attention is so rarely concentrated on some given object, that the routine which, to a busy man, is hardly noticeable, fills the life of the mind less occupied. The attention is never concentrated enough to lose sight of the successive circumstances which mark the hourly flight of time.

Contrast with the quiet life, the hurry and bustle of a busy one. The dull tones indeed are not absent; they blend into a deep diapason. But above and beyond them, there is a variety of incident which heightens the tension of the life, and in so doing, changes its general harmony in two directions. Vivid and rapid experiences make deep and repeated marks in consciousness, and the attention is too engrossed thereby to take note of anything but its immediate object; the hour-marks of time thus pass unheeded. While, in the second place, each stretch of time raises a barrier of new experiences between the observer and the rest of his past. And so, perhaps, the paradox can be explained, that the past moves slowest away when the present moves quickest.

CHAPTER V.

MEMORY.

46. What Memory is.—Memory consists in the revival of past states of mind; each train of impressions or of ideas leaves in our minds a tendency to pass through the same states again. This revival is usually accompanied by the consciousness that we have experienced similar states before; that is to say, together with the revived elements, there come to mind sufficient associated elements to enable us to refer the original impressions to our own past, and if these associated elements are very clear and distinct, we can fix the circumstances of the original occurrences with considerable accuracy. We are aided in recalling the occasion of some particular experience if we dwell upon its details; some train of thought is at last hit upon which conducts us to well-determined points in our memory, up to and from which we measure the lapse of time.

We may mark off three stages in this recognition that some train of thought answers to a past experience of our own. Firstly, complete uncertainty. Some people will tell you the same anecdote time after time, and only be haunted by a suspicion (which they at once dismiss) that they have told it you already. The fact that they told it to *you* has

MEMORY.

not impressed itself on them. Somewhat similarly, a man will come across some striking expression, thought, or other source of suggestion, and forgetting the source of his inspiration, will retail it as his own. Secondly, we recognize memories as our own, as referring to our own experience, and yet are unable to localize them at once, and we succeed after an effort of the attention. This process, by which at last we attach a particular idea to connected ideas, is called *recollection*. The remarkable way in which witnesses in a court of law can recall the whole of the occurrences of a given day in a distant past, and their order, occurrences which apart from this exercise of recollection would have continued to fade from their memory, speaks eloquently for the power of recollection, and for the skill of the legal profession in setting it to work. Thirdly, some events and experiences stand out clear and unforgotten amid hazier outlines, and the smallest suggestion brings them back undimmed with a crowd of determining circumstances by which we can fix their occasion.

We have regarded trains of ideas as the reflections of our past experiences, and as if the order of our ideas were necessarily the same as that of some past experience. This is not always the case. In the discipline of school and business, the order in which each new acquisition is made, matters less, and tends to be less remembered, than its place in the system of ideas to which it belongs. Here then we no longer refer ideas to the order in which we first had them, but to an order in which perhaps they never presented themselves to us! How is this possible? The answer is this. Ideas are not merely the *materials*, they are also, as Locke points out, the *instruments* of knowledge.[1]

[1] *Human Understanding*, Book ii., c. 33.

Ideas tend to blend and to combine one with another, while they refuse to enter into combination with yet other ideas; and the first order of experience is scarcely to be recognized in the form into which by this means our acquisitions are recast. Here we have an important instance of the activity of attention. By contemplating the materials which the memory is constantly presenting, and by passing to and fro among them on paths already marked out in them, or suggested to us by foreign points of view, it brings them into groups connected by inner bonds of some sort or another—likeness, subservience to the same end, origin in some common cause and so forth.

And attention, which performs these higher duties, can also emphasize the mere order of the first impressions; trains of memories of insufficient strength may be greatly improved by passing along them time after time. But a memory which merely reflects the past is a sign of deficient mental power. Mrs. Somerville tells the tale of an idiot who could repeat a sermon verbatim after hearing it once, and did not fail to indicate where the preacher coughed or blew his nose.

47. **Suggestion: Experiments.**—Each revival of a past state takes its rise in some element of the state through which we are passing at the time of its revival, and this process is called suggestion. Like other mental processes, suggestion requires time for its accomplishment. A train of ideas does not always rise clearly to view directly upon the operation of a corresponding suggestion; we often have vague premonitions of ideas which presently shall dominate us, without at once being clearly conscious of them.

Any idea which has on previous occasions been followed or accompanied by other ideas, is qualified to suggest these other ideas. Thus if the idea A has at any time been immediately followed by the idea B, then A tends to suggest B; likewise, if the idea B has been conjoined with another idea C, then whenever B comes, C will tend to come too. We may find an example of the former, or *serial* suggestion, in the notes of an air; each note in a familiar piece suggests the next. There is *simultaneous* suggestion at work between the words and the music of a song; each note is entwined with the answering word.

Suggestion, then, acts along lines already marked out by previous experience, and the laws of association set forth its manner of operation. These laws are as follows: ideas become associated through likeness, unlikeness, and contiguity. In association by *likeness* one idea suggests another with a strength proportioned to the number of their common elements. As the likenesses between two objects increase, so does the thought of one tend to call up the thought of the other. This is the logical kind of association; we ought to think of things together so far as they are alike.

In association by *contrast* the contrasted ideas do not suggest one another immediately and directly. What happens is this: In passing from one idea to another we have a feeling of more or less strength in proportion as we are unprepared to receive the new idea. In this way we have impressions of relation between our ideas (§ 31). If such an impression of relation is very strong, it is easily revived and brings up the connected ideas along with it. We remember the shock, and then we remember its circumstances. As we become more familiar with the circumstances, the feeling of contrast becomes less and less, and, at last, passes away. The adult does not think of day and then of night because of the contrast of light and dark, although the young child may; he passes from one thought to another just as he would think of any two things which follow one another. Association by contrast thus passes into the association by *contiguity* or coincidence which is described next.

Ideas which we have had side by side, or in succession, tend to suggest one another. This is the fundamental kind of suggestion. Association by likeness is made up of many associations by contiguity. One tree tends to suggest another tree, because each attribute of the tree, such as having trunk, branches, leaves, flowers, or fruit, is connected by frequent experience with all the other attributes. We usually confine the term " contiguous association," to cases in which suggestion acts along but one or two paths; while those cases in which it acts along many paths are called associations by likeness.

For the sake of brevity I have used the term idea for all cases of suggestion; but this does not cover the whole ground. The suggesting elements may be either impressions or ideas. When impressions suggest the ideas of impressions previously connected with them, as when the scent of a rose brings to mind its form and colour, we have the most familiar form of perception. Each class of sensation—sight, touch, smell, hearing, taste, and muscular sense—is capable of thus suggesting the ideas of connected sensations, and so each impression suggests more than it directly brings. Muscular associations are especially powerful; they co-operate with touch, sight, and hearing, measuring the order and rapidity in which we receive impressions from eye, ear, and different parts of the skin. A weak suggestion may be much reinforced through them; if we have forgotten where we have put an object, the act of putting it away, and its circumstances, may be recalled by rehearsing the actions performed about the same time.

When ideas are suggested by impressions, the process is called *external suggestion*. When ideas are suggested by other ideas, the process is called *internal suggestion*. The former is the usual one in children's minds; only as the years pass and the resources of memory consequently increase, does internal suggestion become more powerful. Ordinarily, the two kinds of suggestion give, and take, place at short intervals; now the world without, now that within, dominates us.

The processes of suggestion have been investigated in two directions; as to the time they take, and as to their nature. Mrs. Bryant and Prof. Cattell made the following experiments:[1] They put lists of nouns (concrete, abstract, and proper), of verbs and of adjectives before the persons experimented on, and asked them to name the things which they suggested to them. The time taken in this process of association was from one to seven seconds. This difference was due to varying readiness in conforming to the conditions of the experiment, and of course to varying quickness of mind. Thus the process was much more speedily performed by the higher classes of a school than by the lower ones. The contents of the memory and readiness to decide determined the variations in the time taken no less than the speed of the process of suggestion itself. Thought, then, is not so swift as some have said.

The most frequent lines taken by suggestion were as follows: From part to whole, and from whole to part (*e.g.* house to street, and house to room); here is simultaneous association. Forwards and backwards (*e.g.* house-top, glass-

[1] *Mind*, vol. xiv. p. 230.

house); here is successive association. These forms of association simply reflect the original impressions, and may be marked off as reproductive. The following examples illustrate associations based on comparison and inference, which may be marked off as logical: house-building, here is generalization; house-bricks, here is association of means and end.

48. **The Nature of Revived States.**—We have seen how memory depends upon and refers to past experiences, and how by the operation of suggestion, these experiences are brought to life again; it now remains to examine the nature of these revivals somewhat more closely.

The revival or idea of an impression of sense, is under ordinary conditions so much weaker than the original impression, that the difference in degree becomes one of kind, and we do not confuse the two. Sometimes, however, and this is especially the case with auditory sensations, we are not quite certain whether we received a new impression, or merely remembered an old one. We often think we hear our name called without it being actually so. In patients subject to hypnotic suggestion, this confusion becomes complete; they cower before imaginary tigers, relish imaginary dishes, and smell imaginary flowers. Dreams also illustrate the manner in which ideas are confused with the impressions of reality.

When we turn from the revival of impressions to that of ideas, we no longer find that the revived elements are necessarily weaker than the original ones. By some flash of insight we gain a clue to an intricate problem, we recur to it time after time, and follow it out further and further. In such a case as this, the later trains of ideas are fuller and stronger than those which they reproduce. And generally we can observe, while watching the current of our thoughts over long periods, how some trains of thought steal in, and gradually gaining strength, dominate the rest, until they are subdued in turn by some new-comer, which was as weak in its first beginnings as they.

Up to the time of the actual revival of a past state, all that exists in the mind is a tendency to go through that state, and suggestion has merely the duty of setting that tendency to work. Now the same element (idea or impression) may have suggestive force in several directions, so that the tendency caused by one experience may be complicated with other tendencies, all starting from the same suggestion. My thought of a large hall or theatre suggests performances which have taken place there, but none of these may have so impressed me that I follow it out to the exclusion of the others, and consequently my memory does not reflect any single occurrence. In the same way our thoughts of a friend do not represent him as we knew him at a particular season; memories of many meetings and conversations blend together. Hence memory does not reproduce the past literally and slavishly, but in mass and representatively. The way in which this takes place may be illustrated by composite photographs in which, through superposing the portraits of many individuals, generic images of certain classes of persons are obtained. The resulting pictures give a hazy background with the features standing out more or less distinctly.

The revival may take the form of a cluster of ideas, or of a series of ideas. We recall simultaneously the different parts of a landscape, while the notes of a melody are remembered one after another. These two kinds of revival may be blended, and indeed usually are. Clusters of simultaneous ideas each tend more or less to suggest new series of thought, while each term in a series has lateral connections with other elements which have occurred simultaneously with it; the notes of a melody not only suggest the next notes, but also come to mind simultaneously with the answering words.

The form of revival is determined in great part by previous experiences, and therefore by the distribution of the objects which gave rise to them. We gradually form clusters and series of ideas corresponding to the objects in our surroundings, which represent those objects in our thoughts. These masses of ideas are called *representative images;* they do not mirror our particular experiences of things, but sum up the effects of many experiences. They may follow upon many sets of impressions received from single objects, such as the theatre or large hall referred to above, or they may come from classes of objects—horses, tables, clouds. In either case the process is essentially the same, similar sets of impressions leave traces which blend together. These representative images are the material from which our concepts are built up, or rather they are concepts in an earlier and undeveloped stage; they have, therefore, sometimes been called *crude concepts.*

When the term image is used, it must not be limited to visual sensations and their ideas; an image, in psychology, includes the ideas of *all* kinds of sensations, and also the motor elements referred to in § 30.

49. **Surroundings as Influencing Memory.**—The contents of our memory depend in great measure upon our surroundings. Images, unless constantly supported by fresh impressions, become weaker and decay. While those which are constantly assimilating fresh impressions, remain vivid and

powerful. To some extent, then, we are at the mercy of our surroundings; the houses in which we live, the people we meet, the books or newspapers we read, are helping ceaselessly in the formation of new clusters of ideas, or in the dissolution of old ones. But the ideas so gained do not merely take their place side by side with the other contents of our mind; they are also active forces moulding our whole mental nature to themselves.

The objects in our surroundings thus help to mould our characters; for by determining what impressions we shall habitually receive, they also fix our habits of memory.

50. **Kinds of Memory.**—Memory is made up of tendencies to remember; it should not therefore be regarded as a single indivisible faculty of the mind, but rather as a bundle of powers. There are as many kinds of memory as there are classes of things to remember. One memory is concerned with forms and colours, another with musical tones, and so on for all the different classes of sensation. This view of memory is supported by pathological observations. Side by side with cases in which the memory as a whole is decayed, cases of partial loss of memory occur. The memory of words or numbers or faces may vanish for a time or altogether. Sometimes the loss may involve some quite particular set of acquisitions. Dr. Beattie relates that one of his friends having received a blow on the head, lost all his knowledge of Greek, though his memory was otherwise unimpaired.[1] This subdivision of memory further appears in the often-noticed fact that each individual remembers some things more easily than others. Rousseau declares that he had no verbal

[1] Ribot, *Diseases of the Memory*, p. 144.

memory, and was never able in his life to learn six verses by heart.[1]

51. The Course of our Ideas.—In ordinary circumstances, the changing impressions made upon us by external things, and the deeper, if less obvious, changes in our general physical state, are constantly bringing fresh trains of association into play, and so overlay the development of our ideas according to their own laws (§ 37). Yet it is conceivable that a series of sensations having once been presented to our mind, all communication with the outer world should cease. In such a case, the ideas of the impressions which were received when the mind was still perceptive of its surroundings, would for the future enter into combination with each other, and thus the world of thought would have its contents exclusively determined by the laws of association. These conditions are almost fulfilled in the ideal state pictured by the *Poet at the Breakfast-table* when he asks: "Build me an oval with smooth translucent walls, and put me in the centre of it with Newton's *Principia*, or Kant's *Kritik*, and I think I shall develop an eye for an equation, as you call it, and a capacity for an abstraction." It has already been pointed out that under ordinary circumstances manifold tendencies to follow out trains of ideas are present to the mind at one and the same time. Let us consider how these would stand to one another in the imaginary case just described.

Some of the series of ideas will have their origin in the same idea; the idea of a hand suggests writing, grasping, and many other actions. Other series, again, will terminate in the same idea; the idea of three added to three, or four added to two, alike lead to the idea of six. In yet other

[1] *Confessions*, Book iii.

series, the middle term or terms may be the same. Our thoughts of Hanoverian and English history are two series of ideas, which flow apart down to the accession of George I., and after running together for a time, fall asunder again with the beginning of the present reign.

Those series which have their initial parts in common may be called *divergent;* those of which the middle members are in common may be said to *intersect;* while those which end in a common member may be called *convergent.*

Summing up, then, we should find in a current of ideas undisturbed from without, an entangled web, so to speak, of series which converge, diverge, and intersect continually; each idea as it comes into the mind being the starting-point of new trains of ideas, or forming part of new clusters.

52. **Memory as Survival of the Fittest.**—The ideas which at any given instant can come to us are limited in number, and in consequence there is, so to speak, a struggle for life on the part of the groups or trains of associations which, taken together, constitute our memory.

How is this struggle determined? By the amount and force of the various suggestions which tend to bring up the competing trains of ideas (§ 38). Let us see how the various kinds of suggestions contribute to the result.

Ideas support one another when there are elements in common between them, as when we look at two photographs of the same place taken from different points of view, or when we read of two individuals whose careers had much in common, as, for instance, Julius Cæsar and Napoleon. In each of these cases the mere association of two groups or trains of ideas, with elements in common, increases the power of them both over us. Our ideas of Cæsar and

Napoleon are stronger, knowing both series of ideas than knowing one alone. It is thus that our ideas of the more familiar objects we come across become so strong; through repeated experiences of a similar kind the common elements *blend*, forming a nucleus, so to speak, round which there gather from time to time constantly different elements. It is thus, for example, that our notions of the different species of animals—horse, dog, sheep—gradually attain fullness, clearness, and strength.

Trains of ideas, in the second place, become strong by contiguous association. This is what Locke charges, under the title of "the association of ideas," with causing unreasonable antipathies and ungrounded prejudices.[1] To be guided by the mere contiguous association of ideas when there is no real community between them, is indeed natural if this chance association falls in with our prevailing customs of life and thinking. It accounts for the implicit and unreflecting assent we give to assertions made by certain individuals or periodicals, or in certain books. It is the spring of half the prejudices which stand in the way of the proper exercise of our reason about the common things of life.

We have traced the essential and accidental ways in which ideas support one another; we shall find that they oppose one another in two corresponding ways. Some ideas are necessarily opposed to one another; that is to say, they cannot be clearly presented to the mind at one and the same time. We cannot think at one and the same moment that a particular object is both blue and red, nor can we think the same statement to be both true and false in the same meaning. And this opposition which holds

[1] *Human Understanding*, Book ii., c. 33.

good between single ideas, also holds good between trains of ideas. Some lines of thought, if often followed out, make us unable to follow out those of another kind. Mr. Darwin found that his habits of exact observation and ordered reasoning concerning natural objects dulled his taste for the creations of art. Even in this case, however, it is possible that the reason lay rather in the want of exercise on the part of the imagination; one train of ideas merely left no room for another train, and there may have been no essential antagonism between them. The opposition is less to be contested in cases such as the dependence of moral insight upon conduct.

Ideas are opposed by others with an antagonism which is purely accidental, when consciousness is so filled by one set that there is no scope for the others to come before us. At each moment there is this neutral antagonism between the ideas in our minds and all others, which are possible to us through past experience or reflection. We are now preoccupied in one way now in another way, by business, by this or that hobby, and in this way whole sets of acquisitions go by the board.

Thus we forget ideas which are seldom brought to mind. They are like stationary individuals who are separated from the main body, not by any backward movement of their own, but by the general onward movement of the rest. Add to this the diminishing strength of ideas of sense impressions, and we can understand why our memories of actual sensible objects, the vanished hand, the silent voice, are so elusive. As we have seen, ideas are not so transitory as impressions and their immediate effects (§ 48).

Since our power of retaining and recalling ideas is limited, it follows that a good memory implies the power of forgetting. We must select, reject, and ignore, as occasion demands, in order that we may bring our whole forces to bear on each case in point. To have the mind full of vagrant associations which leave no room for pertinent memories to arise, is to be encumbered not equipped. "You offer me an art of memory," said Themistocles, "rather give me one of forgetfulness."

53. **Imagination as due to Activity of Ideas.**—Imagination is sometimes opposed to memory, as if the mind were exercising a different function in each case. In reality they are different stages in the same process; a revival of a past experience is not a mere reproduction, it is to some extent a transformation.

The mind being equipped with numerous clusters of ideas, each ranged round a nucleus of common elements, a fresh experience, if at all familiar, is superposed, so to speak, upon the answering representative image of other similar experiences. We rarely are impressed at a single occasion with all the characters of an object or of an occurrence, and we fill out the gaps from previous knowledge. Seeing a patch of blue, we think of an iris; or a raised arm suggests the brawl that will follow. What we call imagination is the power of thus following out a whole train or cluster of ideas on very slight suggestion. In memory pure and simple, this following out usually answers more or less closely to some actual event or object, but still the process is similar. Imagination becomes more marked when the mere idea, and not the actual impression, is enough to set our minds to work.

We have spoken hitherto as if the suggestion, whether idea or impression, actually recalled the associated ideas. This is the superficial way of regarding the matter. In reality, some system of memories being present to the mind, meets the suggestion half way. Every activity of the attention, every clear and distinct perception, implies a certain pre-adjustment of the mind to the impression or idea received, and this pre-adjustment consists in the presence of a large mass of associations with which the new impression or idea may unite. Thus each mental operation implies an approximation between the new experience and the old memory. Each new acquaintance, for example, is pictured by our minds at first on the lines of the human nature we are already acquainted with; and our intercourse with him is interpreted in the light of that pre-conception. At the same time his conduct will modify our ideas of human nature, and will thus change our attitude to every further person we meet. (See note, p. 122.)

It follows, then, that our minds exercise imagination at nearly every step they take. The only essential difference between my imagining how the other party to a business transaction will act, and a novelist delineating a character, is one of degree. The latter makes an effort which is more sustained, implies a greater divergence from actual circumstances, and perhaps starts from less suggestion.

54. Habits of Memory and Constructive Imagination.— Memory is itself a habit in the making; to remember is to repeat a previous state. But there is a sense in which we can speak of habits of memory. Those trains of ideas to which we constantly recur, and on which we dwell with special emphasis, may be marked off from the rest. We have seen how these are determined in part by our surroundings; they are even more determined by our main interests and aims. We constantly follow out the ideas which lead up to, and start from, those events in the past or those possibilities in the future, which have affected, or promise to affect, the main current of our life. This depends on memory throughout, for even our forecasts are based on previous experience. The expectation of the future is the memory of the past. According as we surrender ourselves to the answering trains of memories or no, such events and possibilities acquire a greater or less prominence in our mental landscape. The best check upon the habits of reverie and brooding, which indulgence of the memory begets, is to remain in constant intercourse with the actual objects and business of life. The memory leads the mind into very perilous paths indeed, when it plays idly and wantonly round past indulgences, or self-assertions at the expense of others, and "grows half guilty in its thoughts again."

Of course our ideas of future events can only be formed by exercising the imagination; but when those ideas have once been formed they can be revived in the same way as other ideas.

The reorganizing of ideas into new combinations, no less than their mere reproduction, is determined by our interests. Our habitual imaginations are as self-centred as our memories. The chief shape taken by this process is the imagination of ourselves as we shall be affected by future circumstances. By dwelling on some future state to be sought or to be avoided by us, we gradually obtain ideas of the circumstances which may be expected to lead up to, or away from, such states. In this way our aims become definite; and in proportion as the circumstances which lead up to them are within our power, practicable. In childhood the desired state is often very far removed from possibility, in a few years it becomes the more rational vision of the young man, until at last the power of imagination decays, and the old man is satisfied with food and a warm fireside. It is only the finer minds whose glance takes in wide horizons, and who find their individual aims merging into the one general hope of human progress. But they have their compensations for this loss of self. Habits of imagination which grow out of, and depend upon, the crude ideas of personal indulgence and deprivation, become faint when the body, which is the organ of those states, begins to lose its first responsiveness to external stimulus. While the aims which seek to be realized in the higher forms of service, and even in self-culture and business, provide mental stimulus to the end.

Let us now proceed to draw one or two inferences. One of the most important exercises of the imagination is the response to symbols. Symbols are merely the means of suggestion; hence they are only of use when the person to whom they are addressed has the key to their meaning in his own trains of ideas. Words form a striking example. With the exception of a few words which suggest by their sound the objects denoted by them, words are purely symbolic. Another use of symbols is seen in artistic creations; the persons to whom they are submitted are always called upon to fill out the blanks due to the limitations of material, treatment, or subject. One warning is necessary; we must distinguish between our following out an artist's suggestions and realizing the ideal he wishes to portray, and the act of mind by which that ideal was originally conceived by him. We are not artists merely because we can understand and appreciate works of art. Nor are we thinkers because we can trace the path taken by the thoughts of another.

Through the blending of ideas which have elements in common and the consequent clustering of some ideas, and the isolation of other ideas, the traces left by our experiences are reorganized. Each act of contemplation by which two sets of ideas, received it may be at very different times and places, are brought together, strengthens their affinity if they have common elements, and marks their opposition if they have not. Hence the quiet hours in which we take stock of our acquisitions and our losses, are not less fruitful for our mental progress, are in truth more fruitful than the mere continued assimilation of new impressions. At such times tendencies which have gathered strength unsurveyed by the attention, appear in their full power and are brought to judgment. At the same time the fundamental systems of ideas exert their attractions and repulsions on each such train of memories that comes up for review. Our conceptions of duty, self-interest, public weal, must assimilate to themselves each new experience, or else become conformed to it.

This aspect of ideas will be considered more fully when we consider the nature of willing. Perhaps, however, we may here remark the profound psychological insight which underlies many religious phrases, hackneyed indeed though they may have become. In appealing to the power which certain conceptions of life and conduct exercise in the mental history, assimilating, transforming, expelling, religious teachers are on the firm ground of experience.

Note.—The process by which a psychical system takes up into itself a new experience is called *apperception*.

CHAPTER VI.

REASONING.

55. Beginnings of Reasoning.—The power of reasoning does not depend upon a knowledge of logic. "God has not been so sparing to men to make them barely two-legged creatures, and left it to Aristotle to make them rational."[1] When Locke wrote these words, he was doubtless thinking of the exaggerated importance attached by the schoolmen to the syllogistic method; the Greek thinker does not deserve the implied reproach; his system stands in closer connection with the ordinary processes of thought, than we should be likely to infer from the form in which it is usually stated now. All that he claimed to do, was to bring to light the principles which underlie the processes of thought; he knew that it makes very little difference in our thinking whether we know the names and kinds of those principles or no.

The path of reason begins to be prepared as soon as, amid the chaos of successive and simultaneous sense-impressions and the traces left by them, definite clusters and series of associations are formed. The first form taken by these clusters is that of representative images (§ 48); each such image in a manner stands for and represents a class of

[1] *Human Understanding,* Book iv. c. 17.

objects. Very often some single object is taken to stand for all its class, and we couple with its image the reflection that it must be taken as typical only so far as it is like the other members of the same class. Thus each figure in Euclid is taken as a type of all the others of the same kind. By reference to such representative or typical images, our thought extends its application over many objects at once. Our power of general reasoning depends on being equipped with a store of experiences, organized into groups and expressed in representative images.

Viewing the trains of association as a whole, we shall find them to start from various centres, and to form systems of memories of more or less complexity; for example, those denoted by the words home, business, religion. Each of these great systems is made up of smaller systems. Thus the following are some of the groups of associations which enter into the whole, denoted by the name business; correspondence, division of labour, and all the various elements entering into business organization; all the conditions which determine market prices with their consequent effects on both demand and supply; our plans as to our future career in business. We might proceed further and analyze each of these subordinate systems into yet others; thus the state of the market may be determined by, and regarded by us in the light of, agricultural conditions, the supply of labour, the state of the money market, and so on. By carrying this analysis sufficiently far, we should at last arrive at the single experiences or their representative images on which these structures ultimately rest. Our memory thus supplies material within the limits of which we exercise discernment and reflection. (See note, p. 147.)

56. The Nature of Concepts.[1]—When the attention is first directed to any cluster of memories, say of some strange animal seen for the first time, they pass before us and leave us with a resulting general idea, which is usually so vague that we cannot recognize its character and mark it off. This is our ordinary experience when we begin to reflect, and just because our ideas work so vaguely we, at times, are incredulous as to our own powers of reflection. At other times our minds are quick to seize the salient and characteristic marks of a train or cluster of ideas, and we are so impressed that we can recall them as a standard by which to test similar future experiences. Very often a second or third, or indeed, many repetitions, of a train or cluster of ideas, is needful that we may receive this characteristic impression.

As soon as a cluster or series of ideas becomes strong enough thus to occupy the attention, the ways in which its various elements stand to one another, become clear. We can understand this easily when it is merely a question of forming a concept of a class of similar objects. Thus, if trees be taken as an example, the representative image of a tree forms the basis of the answering concept, and includes the possession of trunk, branches, leaves, flowers, fruit, and so on, as well as ordinarily the putting forth and the shedding of leaves in due season. But observe, such a representative image as that of a tree, only becomes a concept when we associate with its constituent elements the ideas of the relations between them. We have to take into account the ways in which trunk, leaves, flowers, and fruit determine one another, and how they are related to the objects in their surroundings.

[1] Conception is the name of a mental process, concept of a mental product.

When, however, we ascend to the larger classes, such as plant or animal, formed by the assemblage of the smaller classes, we can no longer represent them to ourselves in the form of a single typical image, we must resort to symbols. It is no longer a case of reproducing actual impressions, but of remembering the ways in which the impressions caused by those objects must stand in order to justify us in giving them the class name. Thus, the concept answering to the name animal includes the several characteristic attributes,—nutrition, waste, sense and mobility—which usually attach to animals. Summing up, then, we shall understand by a concept *an assemblage of more or less complex ideas grouped together in a certain proportion and manner.*

57. **Concepts and Naming.**—The process by which representative images and, ultimately concepts, are formed, is much quickened by the use of language. Likenesses which would have escaped the child's undeveloped powers of observation and reflection, are suggested to him, and guaranteed, when he finds the same name applied to objects or events which resemble one another. Words form meeting-points for clusters and series of ideas; and often in the child's experience, words are at first the single points of community between groups of ideas. This tie of word and object soon becomes very strong for those things which enter into daily life, and are therefore often mentioned. Now the child sees an object and recalls its name; now it hears its name and recalls the object. This process of verbal suggestion goes on side by side with the involuntary association of those objects which are characterized by the possession of common qualities. As the child gathers knowledge, he traces for himself the similarities, by reason

of which he has been taught to apply given names to objects of given kinds.

Thus the process of forming concepts not only depends on the formation of representative images, but also on the right use of words; and the latter is important, since it so often precedes the other method. In order that those objects, which really belong to one class, should be thought of through one concept, the concept name must only be applied to those objects which are really alike. Thus the whale and the seal must not be spoken of as fish, or else they will be confused with objects which are of a different class altogether, such as salmon, eels, dog-fish.

The use of the concept name in the singular is instructive; it generally takes place first when a representative image or symbol has been formed. We speak first of man, or the soul, when we begin to regard all individual instances through a single representative group of ideas.

But when we pass rapidly in thought from one thing to another, there is not time even for the representative image or symbol which accompanies each word, to rise clear to consciousness. We deal with the words as with algebraic symbols, in the knowledge that, if needful, we can interpret each word into the associations it symbolizes and represents.

58. **Clearness and Strength of Concepts.**—The value of a concept depends on the number and variety of the images on which it is based. Those concepts which rest on a narrow experience are unfitted to have a place in reasonings as to affairs which lie beyond it. Further, it is not enough that we may have seen many examples of a class of things; unless those which we have seen include *all* the chief species of the class, the concept will be imperfect. Again, the concept must continually be supported

by fresh images. All mental products tend, as time passes, to become more faint; and it is quite possible to retain habits of expression long after the images on which they were founded at first, have faded from the memory. In order, therefore, that concepts may not only be, but remain, clear and strong, we must remain in contact with actual life, and "accustom ourselves to things themselves."

59. **Psychology and Logic.**—Here we come upon the special subject-matter of logic. Logic investigates those forms of thought which subserve the attainment of the truth, and in our ascent up the scale of mental operations we have now reached the forms in question, the concept, the judgment, the inference. The psychologist is interested in logical and illogical forms of thought alike, for he has to explain them both; the logician, on the other hand, regards mental operations from a special point of view, their value, namely, as a means to attaining the truth, and in appraising this value he applies a standard to which they should conform. The operations, therefore, described in logical hand-books, are not to be regarded as specimens of the ways in which men actually do argue, but as types to which their arguments, so far as they are true, will conform. Logic has but one group of relations answering to each concept-name; psychology has as many as there are individual minds to think the concepts. For no two individuals do the names, house, tree, arouse the same representative images and the same concepts, although the logical concepts answering to tree and house are fixed and permanent.

It will be convenient here to consider the meanings of a few logical terms which frequently occur in connection with the operations we are studying in this chapter.

Any number of things which may be viewed together, form

a *class*. We might think of all the things in the sea as forming one class, and call them by the class-name, marine. Everything in Nottingham might be regarded as forming another class. It is noteworthy that we have not separate class-names for every possible class; only those classes have names, which are very familiar or very important compared with others. When things are viewed from new standpoints, they often require to be named in these connections. The substance of many new works consists in this grouping of familiar things under new descriptions.

To return to the term class. Any class which can be regarded as forming part of a larger class, is called a *species* with respect to that larger class. Thus, horses and dogs are species with respect to the larger class of animals. Any class, on the contrary, which can be divided into smaller classes, is a *genus* or *kind* with respect to those classes. Thus, the class animal is a genus or kind with respect to the classes composing it—horses, dogs, &c. Since one and the same class may be regarded now as forming part of a larger class, now as subdivided into smaller classes, it may be a species from one point of view, a genus or kind from another. It will be a species with respect to the class larger than itself, a genus with respect to the class smaller than itself. The class of quadrupeds is a genus with respect to the classes horses, dogs, &c.; it is a species with respect to the larger class, animals, in which it is included.

A *general* assertion, then, is one which applies to a genus or comprehensive class of individuals. A general assertion about animals is one which applies to all the genus, and therefore to all the constituent smaller classes or species, such as horses, dogs. A *special* assertion, on the other hand, is one which is exclusively applied to one

K

species of a larger class, and not to the other species of the same class. Thus, a special assertion about dogs applies to them, to the exclusion of foxes, wolves.

The same distinction is pointed at in the pair of terms, universal and particular. A *universal* assertion covers all instances within the sphere to which reference is made. A *particular* assertion applies to one or more, but not all, the instances of a given kind. Thus, all men are mortal, is a universal assertion with respect to the class men; some men have dark skins, is a particular assertion. General and special, then, are terms referring to the grouping together of different classes; while universal and particular have reference to the individuals which compose a class.

Concrete and abstract are another pair of opposed terms. *Concrete* names are names of things which can be perceived through the senses; *abstract* names refer to the qualities of things, and to relations between them. Thus, desk, river, tree, are concrete names; so also are adjectives like white and red, for they are the names of white or red things. On the other hand, whiteness and redness are names, not of things, but of attributes; they are therefore to be called abstract names. Whiteness and redness never occur by themselves, apart from other qualities; if we wish to think of them separately, we must "draw them away" from the connected qualities. We must abstract, for example, the quality white or red from the other qualities (texture, weight, smell, etc.) of an apple or a billiard ball. This is fixing our attention on part of a percept or an image of a thing, and not upon the whole.

60. **Logical Concepts.**—The functions of concepts in logic are chiefly the two following; they enter into logical

judgments and into logical classification. Now, both judging and classifying require that the things which are classified, and about which judgments are made, shall have their characteristics well marked off, so that we know exactly what we are talking about, or classifying, and thereby avoid the risk of confusing them with other things. A logical concept will consist therefore of a group of ideas answering to the characteristics of the objects to which it applies, and connected together in a certain manner and proportion. This ideal concept may almost be thought of as a diagram or formula,[1] to which the notions acquired in our individual experience must conform.

The fitness of a concept to be used in ordered reasoning, depends on the exactness with which it answers to all the objects included under the concept-name and to no others. That is to say, the marks or attributes included in the concept must be such as to exclude all objects other than those of the class in question. Our concept of fish must be such as to refuse admittance to creatures like seals and whales. The mere fact of living in the water is not sufficient to cover the class fish, and no others; some other mark must be added, say the breathing through gills. If we take up this mark into our concept, we shall be in no danger of mistaking whales and seals for fish. (My lexicon, therefore, which defines fish as "an animal which lives in the water," is not full enough.) At the same time, different classes shade off into one another, and it is very hard sometimes to find characteristic marks which shall perform the duty just sketched. Thus, the African mudfish and the ceratodus, while possessing the general character of fish, breathe through lungs, as well as gills.

[1] Lotze, *Logic*, § 117.

The process of forming a concept passes through three stages:—

Concepts take their rise in clusters and trains of ideas which possess elements in common. The points of likeness rise into prominence and form the nucleus of the concept. In our thoughts about fish, the two attributes, of life in the water and breathing through gills, are among those marks which occur most frequently to us. This stage may be called *comparison;* points of likeness are brought out.

In the second place, we fix our attention on these points of likeness, and turn away from accidental circumstances. In this case weight, and length, which do not mark fishes off from other things, are accidental circumstances. This is called *abstraction.*

Thirdly, the group of ideas so assembled become capable of being applied to all objects of the class in question. This application of a single thought or group of thoughts to many things is called *generalization.* Theoretically the resulting group of ideas should not be coloured by association with particular objects. But as a matter of fact, we often take an individual as typical of his whole class, and argue about him as standing for all his fellows. The effect of a recent and vivid experience is always to overlay the representative image with such a typical image, until the latter grows fainter and merges into the representative image, like the last face superposed on a composite portrait.

61. Classification of Concepts.—Every resemblance of whatever kind, which distinguishes a class of objects, is sufficient to justify their being grouped together. Thus, the single property of strong cohesion marks off solid from other bodies. Sometimes the likenesses which run through a class of objects may be very numerous; the properties which constitute our common human nature afford an endless theme for speculation. We may classify the concepts which set forth these similarities, according as these similarities are few or many. At one end of the scale we shall have concepts, the content of which is exhausted by one or two attributes or marks, such as roundness or impenetrability. These are the more abstract concepts; they imply a separation in thought of the one or two marks included by them, from all the other qualities of the bodies in which they are perceived. The concepts become more concrete, as the marks they include approach more nearly to the full number characterizing actual concrete objects—horse, dog, tree, for example. As our concepts within a given sphere of objects become more full, they are capable of being applied to fewer and fewer objects. Thus, if we add the mark of being a metal to the mark solid, we exclude all solid substances which are non-metallic. Generally we may express this by saying that the scope of a concept varies inversely with the content; or in less technical language, the number of individuals to which a concept applies is smaller, when the concept includes more marks, and larger when it includes fewer marks.

On this principle, the points on which many persons are agreed are often so few in number, that they scarcely make it worth while to obtain the expression of agreement. The ideas held by all minds in the same meaning and application are few and unimportant, if, indeed, such exist.

When we pass from the concrete, the particular and the special, to the abstract, the universal and the general, we do not altogether lose sight of special characteristics. We must not, indeed, connect with our concept of man, the idea of any particular height, weight, or colour of hair; yet we must remember that every man will have some height, some weight, and some colour of hair. And in passing from the images of particular individuals to the class-concept, the special marks must be taken up into it as so many possibilities, and we must take note of the limits to those possibilities.

When we make the further advance from the class-concept to the general concept under which it falls, the process is still the same. Thus, reason is found in man in a quite unique form. We must not, therefore, include in our concept of animal the degree of reason found in man; yet we must couple with our ideas of the marks which distinguish animals generally the reflection that these general qualities may exist along with human intellect and conscience. The abstractions of logic should not be bare abstractions, but such as can clothe themselves on demand with all the fulness of reality.

62. Function of Definitions.—In order that we may use our concepts in ordered reasoning, they must answer to the individuals included under the class name. But all mental products, and concepts among them, gradually become weaker as time passes, and concepts which may have conformed to the logical ideal at one time may gradually lose distinctness, and even be deprived of some of their component ideas. Here definitions come to our aid; by their means we are enabled to reinstate concepts in their first form, even when we cannot repeat first-hand acquaintance with the things themselves to which the concepts answer. Definitions have also the not less important function of indicating to us, as objects of thought, things which may never have come under our direct notice,

In this way they not only regulate our reproductive associations—they also point out the path by which constructive association may grasp objects which go beyond our previous experience. Such a definition is like a recipe detailing the ingredients of thought to be employed in forming a new concept.

Definitions, it is to be noticed, usually start from some more familiar concept than the one to be defined, and instruct us how we are to modify, extend, or limit it. Explanations and definitions to be learnt by children, should always start from some concepts which are already familiar to them.

63. **Nature and Occasions of Judgments.**—The term judgment has a wider meaning in psychology than in ordinary usage. Whenever the mind attends to its impressions and ideas, and notices how they stand to one another, it performs an act of judgment. It is this process of being impressed with the relation between our experiences, which constitutes the act of judging; to form a judgment about anything is thus the crowning stage in the process of attending to it. Objects of thought are at first dimly presented to the mind; under favouring conditions they rise to prominence and dominate the attention; and lastly, the relation in which they stand is comprehended by us in a judgment. The proposition in which the result of a judgment is expressed, may be interpreted thus—the subject is the thought from which we start, the predicate is the thought to which we connect it, and the copula expresses the form of the connection.

The occasions on which we naturally form judgments, are those only on which the attention is aroused. In other words, they depend on striking changes in the course of our ideas. The child's judgments are always expressed at first in a tone of voice which indicates surprise or interest, and usually both these emotions. Thus, the language in which judgments are expressed is, at root, a form of emotional expression.

When we are not perfectly at home with our company, we fill up the time and relieve our uneasiness with conversation; judgments which do not express any real interest are passed on the weather, on politics, or on literature. In company, Rousseau used to start off babbling words without ideas, only too happy when they bore no meaning at all.[1] It is only in the society of our nearest friends that we can maintain silence. For in circumstances where convention is put away for a time, or does not exist in its civilized form, judgments are only expressed when some thought strikes the mind with a shock of surprise, or has special interest for us. Savages are often thought taciturn, when, having nothing to say, they refrain from saying it.

The thought from which we start in forming a judgment, that is to say, its subject, may be of two kinds. We may be thinking either of the individuals to which the thought applies, or the ideas which it includes; in the judgment, "man is mortal," we may think of the attribute mortal as attaching to all individual men, or as attaching to the other qualities of man generally. A judgment which refers to the individuals included under the concept, is called extensive, since it refers to the extent of the concept. A judgment which refers to the ideas included in the concept is called intensive, since it refers to the intent of the concept.

Extensive judgments may refer to all, or less than all, the individuals included in the subject. When such a judgment refers to all, it is called universal or general;

[1] *Confessions*, Book iii.

when it refers to less than all, it is called particular. Thus, "all men are mortal" is an universal judgment, "some men have dark skins" is a particular one.

Notice that, when all the individuals included under the subject are to be thought of without exception, we may express the subject by the concept-name in the singular; instead of saying "all men are mortal," we may say "man is mortal." The concept, being a compendium, as it were, of all the connected images, can interchange with them.

Intensive judgments are divided into two kinds, according as the idea forming the predicate is, or is not, included in the group of ideas forming the subject. When the predicate is regarded as forming part of the subject, the judgment is called analytic, as "splitting up" the subject. When the predicate is regarded as something outside the group of ideas included in the subject, the judgment is called synthetic, as "putting them together." Any mark, however important in reality, remains outside a concept until it has been connected with it by experience. When a child sees a magnet for the first time, it forms a synthetic judgment that iron can attract; this possibility of attracting is taken up into its concept of iron. But this attribute seems almost self-evident to the advanced student of magnetism, and a judgment expressing the fact is in his case analytic. Thus, for the psychologist, the same judgment may be now analytic, now synthetic. For the logician there is no choice; that iron under given conditions can attract, is for him an analytic judgment in the present state of human knowledge.

64. **Trains of Reasoning.**—In psychology the phrase, trains of reasoning, has a wider meaning than in logic; the psychologist observes *all* the paths by which the mind

reaches a given judgment, the logician confines himself to observing those by which a right judgment may be attained. All trains of association, determined as they are by habit, suggestion, imitation, interest, are as valid in the eyes of the psychologist as the sole logical ground of belief, conformity to actual fact.

The theories of the syllogism and of inductive logic do not, any more than the theory of the concept, actually represent the workings of the mind. They furnish ideals and standards by which we can check our mental operations. Just as the concept furnishes a standard of clusters of ideas, so the syllogism and inductive method furnish standards of the trains or series of our ideas. By tracing the points at which our thoughts diverge from these ideals, we are enabled to detect the origin of any errors into which we may fall.

65. **Belief and Doubt.**—To believe is to tend to form a judgment, and the strength of the belief is manifested in the strength of the tendency. So long as the connection between any two mental elements is not contradicted, it will involve the attitude of belief whenever it is presented to consciousness. Until the mind has been schooled by experience to a more or less critical attitude, it believes nearly everything it hears, and everything it says; it is possible to tell lies, not only so as to be believed by other people, but until we believe them ourselves. The act of believing, however, is limited to those cases which are not strongly at variance with our previous experience. When there is such an antagonism between two of our ideas that we cannot bring them together, we are in the state of disbelief with regard to their connection together. It is on the strength of our incapacity to join ideas in the way suggested to us by travellers and others, that we disbelieve them, a quite natural method, since we *must* judge

on the basis of our present knowledge. A black king, on receiving an account of his travels from a subject who had been to Europe, found it so incredible, that he ordered his informant to be executed for a liar, and yet the poor man had been as accurate as was possible to him. He should have gauged his sovereign's mental digestion, and suited his tale to the capacity of his hearer.

Opposed to these states of belief and disbelief is the state of doubt; we may pass, or refuse to pass, from one idea to another, but the decision one way at once excites us to think the opposite. This state of doubt implies a certain degree of intellectual self-control; the natural tendency is to believe or disbelieve without hesitation, while to doubt implies that we balance contending considerations, and refrain from immediate judgment. The power to conceive alternatives is a comparatively late acquisition. "Every one accustomed to young children," says Darwin, "knows how seldom one can get an answer even to so simple a question as whether a thing is black *or* white; the idea of black or white seems alternately to fill their minds. So it was with the Fuegians."[1]

66. Inference : Deduction.—Inference consists in passing from one or more judgments to other judgments founded on them. Inference is of two kinds—deduction, by which we *apply* general statements to particular cases; induction, by which we draw out the general principles *implied* in particular cases. It is an example of deduction to judge that it will rain to-morrow from the appearance of the clouds to-day. Here we proceed from the judgment, "the sky is lowering," to the judgment, "it will rain to-morrow." Usually the way in which the conclusion (or judgment at which we

[1] *Journal of Researches,* c. x.

arrive) is connected with the ground (or judgment from which we start) is left to be filled in by the hearer. For instance, in the example, we do not explicitly state that a lowering sky generally precedes rain.

In logic, all three judgments are presented in the syllogism, or typical series of judgments. A syllogism, says Aristotle, is an argument in which, something being granted, something further necessarily follows from it. First, we have the general principle (lowering skies precede rain); second, the case we are considering (this is a lowering sky); third, the application of the rule to the case (it precedes rain). The first two judgments in this series are called the premises; the last, the conclusion. Let us denote the subject of the conclusion by S, and the predicate by P. We bring them into connection by connecting them both with a third term, called the middle term; let us denote this by M. Then we may symbolize deduction generally thus—

All M is P,
this S is M;
therefore this S is P.

Thus, deduction is a means of bringing two ideas, or groups of ideas, which are not directly connected, into indirect connection by means of their common relation to a third group. The general principles state what characteristics go together, here, lowering sky or rain; then, whenever one characteristic is found in any case, we go on to think of the other. The lowering sky suggests rain. Or, to put the matter in the form of a rule, we may say with the schoolmen and with Aristotle, "whatever has any characteristic, has that of which it is characteristic."

M has the character P,
S has the character M;
therefore S has the character P.

In other words we justify our asserting a connection between two ideas which is not perfectly obvious, by showing how they are both related to a third idea.

In ordinary conversation we do not trouble to state the general principles; we assume that they are present to our hearer's mind. The effect of daily intercourse and reading is to equip us with a stock of maxims to which we can appeal; and we take them for granted in our conversation with others. They consist chiefly in certain assertions that such and such aims in private life, or in politics, are desirable, and that, therefore, certain courses of conduct are to be followed; in the sphere of religion, certain views as to creed and doctrine are assumed. We know the ground we stand on when talking to familiars; the case is different with strangers. In passing from one judgment to another, the reason which would justify our so passing is not known to be conceded by them. Hence the constraint we feel in a man's company until we have sounded him.

The syllogism does not only suggest to us that we should from time to time state the general principles involved in our arguments; it is also a means of developing the consequences of the maxims which we receive on the authority of others. Taking a general statement, for example, "man is mortal," we bring it to bear on particular instances—this is a man, therefore he is mortal. The case is the same when we make use of a general principle which we have gathered in our experience. The child, finding the fire in the grate burns, thinks that the character of flame and burning go together, and interprets its new principle to the case of the gas, which perhaps is beyond its reach.

The rule which stands at the head of a syllogism is an analytic judgment, and implies that the quality asserted of the subject is included in our idea of it. Mortality is implied in the cluster of ideas denoted by the name man.

We may imagine the mind, on finding some of the marks of humanity, to be carried along to another mark, the idea of mortality. Let $a, b \ldots$ stand for the first marks, and let m stand for mortality; then the mind justifies its association of a, b, and m, together by reference to the articulate scheme of marks, $a, b, c, d \ldots m$, which make up its concept of man. The middle term is thus a group of marks connected together in certain ways, which set forth the lines on which our thoughts may move. By the operation of the attention, the cluster of ideas—the concept man—which determines the particular association—man and mortality—stated in our judgment, is brought to light.

The process of deduction thus implies two things—we must have a stock of maxims or general principles, and we must be able to apply them; we must be able to pass along the trains of thought which the constitution of our general ideas suggests. Our concepts must be so clearly articulated that they can guide us whenever we come across some of the marks included in them. We must be able to catch the true relations of new experiences to old experiences. This act of applying our old knowledge to new cases is a typical example of "understanding." "A Damara never generalizes; he has no name for a river, but a different name for nearly every reach of it. . . . A Damara who knew the road perfectly from A to B, and again from B to C, would have no idea of a straight cut from A to C; he has no map of the country in his mind, but an infinity of local details."[1] Here there is an absence first of the general idea of a river, or of the space relations included under the idea of a triangle ; and there is the consequent inability to grasp the detached circumstances together in the manner typified by the syllogism. The various characteristics of the river do not suggest the general notion of a river in which they are all connected together. From the connection of A and C with B, the relation of A to B is not suggested.

67. **Inference: Induction.**—In induction we draw principles which shall apply generally, from particular instances that have come under our notice. We may represent the general method of induction in this way—this, that and the other instance of a class agree in some particular ; therefore all the instances of the class agree in that particular. This, that and the other man are mortal ; therefore all men are mortal. This, that and the other M is P ; therefore all M is P. The important point in induction is to know how far we may apply what holds good in one or two instances to all other instances of the same kind—to know how far we shall be justified in saying that, because every instance we have seen has some characteristic, therefore all the instances, including those which have not come under our notice, have the same characteristic.

It is by this process of induction that scientific laws are established. This, that and the other metal is an element ; therefore all metals are elements. This, that and the other substance gravitates ; therefore all substances gravitate.

But induction, like all scientific procedure, has its foundation in everyday methods. We are always forming inductions, always generalizing our knowledge, always applying the

[1] Galton, *Travels in South Africa*, c. vi.

result of our observations beyond the scope of the observations themselves. This is done without any conscious effort of thought. Thus, the colour white entering into the child's perception of a swan on every occasion, becomes part of its representative image or concept of a swan. Hence, whenever it thinks of a swan, it thinks of it as white; and if it puts its thoughts into words will say, swans are white. The aptness of this example is rather increased than lessened by the fact that some swans are black; for it is characteristic of this method of thought, that the principles we attain by it being based on our previous experience, may be qualified, or overthrown, by the extension of that experience. But the more wide and diversified our experience is, the less likelihood is there of this contradiction of the old by the new.

To generalize accurately—for this is what induction implies—has two chief conditions. Our experience must have been *wide* and *varied*. It must be varied in order that we may come across the things we are generalizing about, under all manner of circumstances. It must be wide, in order that, by the repetition of experiences of the same kind, they may make an impression upon us; it is by this repeated observation of objects of the same kind, that we come to know them accurately and minutely. While, on the other hand, varied experience, by presenting to us similar objects under different conditions, enables us to distinguish their essential from their accidental characters. The method of Darwin is an illustration; in his inquiry into the variation of the characters of animals and plants, whether in a state of nature or not, he "worked on the Baconian principles, and without any theory collected facts on a wholesale scale." As he continued his studies, points of likeness disclosed themselves between the many diverse examples he collected, and suggested generalizations which should cover all such cases. But he was never content to put forth a theory until he had made repeated observations under all manner of circumstances. After commencing his note books, he waited twenty-one years before publishing his views in the *Origin of Species*. Thus the temper required in making inductions—a temper so eminently displayed by Darwin—is one of patience and self-restraint; patience in gathering wide observations, self-restraint in refraining from coming to a conclusion until we have made observations under all possible conditions.

Of course daily life is well supplied with habits and maxims sufficient to guide us in our somewhat narrow round; it is only rarely that we have the duty of forming a generalization for ourselves forced upon us. Books and human intercourse supply the necessary rules by which to arm, to justify, or to condemn ourselves, and our impressions are easily poured into the moulds provided by surrounding influences. But for all that we sometimes generalize for ourselves, and especially with reference to the characters of those individuals we constantly meet. We have to draw our own conclusions as to them, and as to the way we should act towards them in the business of life. In their case we often jump to conclusions on the ground of a single example. Our first impressions coloured by a single act of generosity or close dealing, of frankness or disingenuousness, of courage or weakness, lead us to infer rightly or wrongly that they will always act in the same way.

68. **Analogy.**—In complete induction we argue from given instances to others which are exactly like them. But even when the likeness is not complete, when it consists in

the possession of but one or two common characteristics, we tend to apply what we know of one case to other cases. "Who drives fat oxen must himself be fat." The lines of suggestion along which the mind moves under stray influences are those of analogy, and the associations which depend on accidental points of resemblance are the basis of analogical reasoning. The force of arguments from analogy increases as the points of likeness become more numerous. Such reasoning has its value in the absence of more full materials for our conclusions. Stray hints drawn from other fields of thought may act as clues for present investigations. Thus Darwin was led to the theory of the survival of the fittest by Malthus' assertion that the population tended to press on the means of subsistence.

This reasoning from analogy becomes scientific, when it is confined to those cases in which the sets of circumstances compared, though not severally alike, yet stand in the same relations to one another. As a rule however the analogies which are so drawn, do not hold good. Politics has always been a soil fruitful in them. Thus Sir Robert Filmer compared the relation between king and subject to that between father and sons, and claimed for the king the father's authority.[1] A similar line of reasoning is thought to justify the claim of the mother country on the allegiance and love of her colonies.

69. **Habits of Reasoning.**—We arrive at most of our conclusions without conscious effort; by repeated experiences of similar kinds our typical images and concepts gain in strength, fulness and clearness, and colour more intensely the impression which each such future experience may bring us. In this way we unconsciously attain notions

[1] *Patriarcha*, c. i. § 10.

and maxims, the grounds of which though lying in our past life, are not always open to our reflection; our idea of an acquaintance is, so to speak, a summary of our previous experience of his character, and yet we should find it hard to justify it in detail.

Since then our trains of reasoning are largely determined by the lines along which our thoughts run, it will follow that the latter will determine the nature of the former. If our thoughts wander idly (being governed by stray suggestion), if they are chiefly fixed on our own interests, if they are drawn hither and thither by conflicting anxieties, we shall either confuse truth with fiction, or turn aside from views which offend us, or hesitate for ever between opposing views.

Note.—The phrase *Universe of Discourse* is applied to the whole group of ideas with reference to which any particular idea is to be understood. This universe may vary in extent from the whole of what is conceivable, to some quite limited sphere of thought. Thus we may be occupied with the universe of chemistry, or even with an entirely imaginary universe, such as that of Norse mythology. By a convenient analogy we may speak of universes of feeling (cf. p. 196), or of universes of desires or ends (cf. p. 222).

CHAPTER VII.

PERCEPTION.

70. Sensation and Perception.—Where the young child feels, the grown man perceives. At the beginning of life, the sensations enumerated in the third chapter succeed one another without apparent connection or meaning. Gradually, however, they leave their traces, and these are built up in the way described in the last two chapters into these complex habits of reflection and imagination, which characterize the mature consciousness. The attitude of such a consciousness to the impressions which it receives, is an index to the transformation which it has undergone.

Our business, then, in this chapter is to show how, by the accumulation of past experiences, and by their organization into clusters and series of connected ideas, we fill out the life of sensation, which is all we have at first, into the later life of perception.

71. Perception and Attention.—It is by the combination of mental elements (impressions and ideas[1]) with one another, that the adult's mind is marked off from that of the child. We may classify these combinations as follows—

I. Combinations of impressions with one another.
II. Combinations of ideas with impressions.
III. Combinations of ideas with ideas.

[1] Impressions = mental elements enumerated in Chap. III.
Ideas = revivals of them arising as described in Chap. V.

PERCEPTION. 149

To receive simultaneous sensations of smell and colour from an orange, is an example of I.; to associate with our visual sensations due to a barrel-organ the idea of the sounds to which it has previously subjected us, is an example of II.; to combine the idea of a boat with that of a stream, is an example of III. The name *presentation* has been given to these complex mental states, in which many elements combined into a group, are presented together.

When any such combination is presented to the mind, and becomes very strong, we are said to attend to it, and its gradual rising to view constitutes the act of attention. As a result of this comes the process of perception: for this is the name given to each act of mind, so far as it is clear and distinct. And as we have already seen (p. 96), an idea in becoming strong also becomes clear and distinct. As I was sketching this afternoon, the wind rose; but busied in my sketch, I clutched my hat with one hand and went on drawing with the other. At last the cold and wind combined made impressions so strong that they compelled my *attention*, and as a result I *perceived* that the wind and cold were connected with the gradual overclouding of the sky. Here the impression caused by the chill wind, was attended to on rising into the clear region of consciousness, and simultaneously the connection of the wind and cold with one another, and with other circumstances, was observed.

The act of perception, then, implies two things—that some mental combination is presented to us, and that we become conscious how its various parts stand to one another, and to the rest of our knowledge. As long as this last condition is obeyed, the name perception is applied sometimes to very faint combinations; attention on the other hand refers to the *strength* of our impressions and ideas.

72. **Perception, Internal and External.**—The name perception is sometimes confined to the combination of ideas with impressions; as, for example, when the sight of an orange is accompanied by the ideas of its taste and smell. But this is unduly to narrow the word. We perceive also when we have combinations of ideas with ideas, or impressions with impressions.

To be conscious of such combinations is the same as to refer the impressions or ideas, of which we are chiefly conscious, to associated impressions or ideas; we refer the colour of an orange to a larger group of sensations, including not only sight, but smell and taste. There is thus a leading element in a percept, upon which the other elements depend. The colour of the orange, the sight of the barrel-organ, the cold impression caused by the wind, are the leading elements in the cases we have just been considering. These leading elements are *simultaneously* associated with connected groups of impressions and ideas. In a percept there is the vivid nucleus and the fainter parts attached to it. As the fainter parts differ in character, so does our interpretation of our impressions change; when Mr. Ruskin mistook the glass roof of a Swiss workshop for an Alp, and then, detecting the illusion, perceived that it was a glass roof, the nucleus of the two percepts was the same, a vision " clear and fair and blue, flashing here and there into silver under the morning sun." It was associated, however, on one occasion, with the ideas of a mountain, on the other, with those of a workshop.

It is by perception that our trains of memories become available at each instant more or less in their entirety. To go through a train of memories is not merely to be conscious of them one after another. But each as it comes, is simultaneously associated with ideas of the preceding and the following members of the series; the preceding ones growing fainter, the following ones becoming stronger, until they in turn take the chief place, only to make way for yet other memories. Here again there is a nucleus of a vivid character accompanied by less vivid elements. The reference of a remembered impression to the occasion on which we experienced it, is also an example of this combination of some more vivid element with less vivid ones. According to the nature of the less vivid elements brought up along with it, will our reference of it to some group of circumstances in our past life be clear or obscure (§ 46).

It is by perception that we perform the various processes of reasoning. Some more vivid impression or idea is always calling up connected elements; they remain in the mind along with it, and its relation to them is grasped. Thus, the nucleus of a percept may remain while we pass from one faint set of associations to others connected with them. Thinking of gold, we have the idea that it is a metal, and this is replaced by the idea that it is an element. We may represent the process thus: *gold* (metal) element. The reader would find it a useful exercise, to denote in a like manner his state of mind with reference to objects from which he is receiving impressions, or of which he is thinking. One's state of mind in making an induction might be denoted thus: *substance* (apple, moon, earth) gravitate.

We may also distinguish our perceptions according to the way in which we regard the elements which cluster round the nucleus of each. There is one very broad distinction which goes right through all our states of mind:

that, namely, which holds between impressions and ideas regarded as part of our life, and the same impressions and ideas regarded as corresponding to changes in our surroundings. We may distinguish in this way between Internal and External Perception.

In Internal Perception each impression and idea is viewed in connection with all the other ideas and impressions which may be present to us at the moment. If we fix our thoughts on our state of mind as such, we gradually become conscious of sense-impressions of various kinds, including those more vague sensations which rarely rise to consciousness. At the same time, all the associations in the light of which we refer our impressions to their causes become faint, and we are left with the mere consciousness of our physical state of the moment. Internal perception, thus pursued, leads us to the bare consciousness of self. It is one of the characteristics of the more intense pleasures and pains, thus to expel from consciousness all ideas but those of the immediate present. To surrender oneself to the impressions of the moment, to give the reins to our ideas, is to be left with this bare self-consciousness, is to strip the mind of the vesture it has woven for itself.

At the other extreme is External Perception. The mind brings its immediate impression or idea into connection with other impressions and ideas, and determines its place in relation to these. Thus, to perceive a sound is to be conscious of it as coming to us from some quarter or other; to perceive the meaning of a thought, is to put it in relation with other thoughts.

In our ordinary moods we waver between these two poles; we are rarely so buried in our immediate state as to be unconscious of our surroundings, or so attentive to our surroundings, or to some train of thoughts, as to be unconscious of our immediate sensations.

From this point of view, to be buried in a train of thought which lies outside of our immediate state, is no less external than to attend to an actual sense impression. To see a picture and to recall it alike imply a certain going out of ourselves. We may classify our results in the form of a table, thus—

Perception, Internal, Consciousness of Self.
             External (a) Sense-perception.
                       (b) Perception of ideas.

73. **Sense-Perception.**—We begin with sense-perception as the most familiar of these three processes. It may consist in the combination of impressions one with another, or of impressions with ideas. As an illustration of the former process, we may take our perception of a circular figure. We regard the various parts which compose such a figure as connected together, without any necessary reference to previous impressions. The more usual kind of sense-perception consists in the combination of ideas of past impressions with immediate impressions. It can only be in the earliest stages of our experience, or under quite peculiar circumstances, that we can receive a group of impressions without their being supplemented by the faint echoes of previous impressions.

The following pages enumerate the chief ways in which our impressions and the ideas of them combine. They are classified according to the channels of sense through which they come, and not according to the objects in our surroundings which give rise to them.

We begin with the combinations which take place within the limits of the same sense, and then proceed to those cases in which impressions and ideas of different senses combine. We then consider how these various combinations are built up into our perceptions of external objects and of our surroundings generally.

74. Combinations of Common Sensations.—The vague sensations which indicate the state of our respiration, circulation, digestion, combine together under ordinary circumstances into a vague presentation of bodily comfort or discomfort as the case may be. In this they are supplemented by the general sensations of muscular freshness or fatigue. All these merge together into one combination. Sometimes the sensations of digestion or respiration rise into prominence, but this is only when the bodily functions are disturbed. Generally speaking, it may be taken that they do not rise from the background of consciousness.

75. Muscular Combinations.—Different muscles may simultaneously give rise to muscular sensations, as when the two arms push against some solid object. We might multiply illustrations of this kind to infinity, by showing how the muscles of the various limbs combine continually in all sorts of motions, and how each combination is accompanied by its own characteristic complex sensation. The ideas of these sensations blend together and form a kind of picture, which is always vaguely present to us, of our muscular equipment. When we have been inactive for some time, the craving for exercise becomes very intense, and each set of muscles reports its restlessness to the mind. The state of fatigue furnishes another state in which we are conscious of sensations from many sets of muscles. Here we have simultaneous impressions combined with one another. But the most important combination is that of present muscular sensations with the ideas of past muscular

PERCEPTION.

sensations. As the arm, for example, is slowly moved through space, the ever-changing tension of the muscles reports itself in a series of muscular sensations passing one into the other. At any given point in the movement there is not only the sensation corresponding to that particular muscular tension, but also the ideas of the sensations which have just been experienced, and of those which will immediately follow. Thus, the perception of the position of any limb consists in part of the muscular sensations answering to that position, and of the ideas of the sensations which have accompanied its movement from its ordinary position of rest; there must further be added the ideas of the sensations which will accompany the subsequent movements.

Most of our movements are produced by outgoing motor impulses which are under our control. Thus, impressions of movement are associated more or less strongly with the consciousness, vivid or faint, of the motor impulses which set them on foot. This element in perception deserves especial notice; from it we get our idea of ourselves as centres from which forces radiate out upon our surroundings. Hence we begin to use the personal pronouns, and to say, "I do this, that, or the other."

If this be true of the one class of muscular sensations, those, namely, of movement, it is still more true of the other class, the sensations of resistance. The outgoing motor impulses, by which we push and pull objects, give a characteristic colouring to the perceptions into which they enter; we become conscious of ourselves as bringing about changes in our surroundings in defiance, so to speak, of their inertia or resistance.

These motor impulses have a share in the production of

most of our sensations; we move the eyes, the hands, etc., and so bring our sense organs under different stimuli from time to time. Thus, through all our perceptions there runs an element of will.

76. **Tactual Combinations.**—Tactual combinations, like muscular ones, may arise between simultaneous sensations, or else between present sensations and the ideas of connected past ones. Whenever any substance comes into contact with several parts of the skin provided with distinct organs of touch, tactual combinations arise; the separate organs of touch cause separate touch sensations which are combined. Thus, in grasping an orange, the various sensations of touch received by way of the fingers combine into the perception of its surface.

Examples of the combination of tactual sensations with the ideas of previous tactual sensations are furnished by the list of composite sensations given (§ 26). Thus, the sensation, or rather perception, of roughness implies the combination of the ideas of past with present sensations, as the reader may discover by passing his hand over his coat, and comparing the result with what he experiences, when he simply lays his hand upon it; the feeling of roughness only becomes clear when the hand has passed over some distance.

Another important class of combination is furnished in cases where two or more neighbouring points in the skin are excited; the sensations thus arising, are generally supplemented by the ideas of those which would be caused by the stimulation of intervening points. Thus, if the ends of a pair of compasses be placed along the bare arm in the direction of its length, ideas (faint it is true) arise of the intervening places. The relative positions of points on the skin are perceived in this way. As the hands or other objects are passed from time to time over the surface of the body, the stimulation of each point is associated with the stimulation of the other points in the neighbourhood. Hence, when two neighbouring points are simultaneously stimulated, the ideas arise most strongly of that series of points which lies between the two points, and therefore has been excited most often, beginning from either of them.

77. **Visual Combinations.**—Here, again, we come across the combination of actual sensations in the cases when different parts of the retina are simultaneously excited. The peculiarity of this case is that not merely one or two points, but whole regions of the sense organ, can be simultaneously excited, and the resulting sensations combined into single perceptions. Thus the clock subtends a very considerable part of my retina as I look up at it. The grouping of those parts of the retinal surface which are simultaneously excited at any one time, determines our visual perception of form.

In the same way as described for touch, the excitation of separate points on the retina arouses the ideas of the excitation of the intervening points. We estimate the relative position of points on the retina, by our consciousness of the excitable points which lie between them.

The more familiar forms (combinations of lines), such as squares, oblongs, circles, ovals, spirals, triangles and so on, are of very frequent occurrence. Thus, if they are partly presented in the field of vision, by the operation of association, we tend to fill in the remainder. This is an instance where visual sensations are supplemented by the ideas of visual sensations.

The fact that the yellow point of the retina must be directed to any point in the visual field of which we wish to get clear vision, is of great importance. If we wish to observe a form with the greatest possible clearness, the several points in its contour must pass successively over the yellow point. Suppose that m stands for the yellow point, l and n for other retinal points, one on each side of m, which successively receive the image of a given point in the contour of the form observed. Then this point subtended the retinal point l before the yellow point m, and after it has passed over the yellow point, will fall on the point n. Thus, side by side with the actual excitation of the point m, are conjoined the ideal excitation of the point l (the corresponding actual sensation being just passed), and also of the point n (the actual sensation being still to come, but clearly inferred from past similar experiences). Thus, generally, the visual perception of form includes the actual group of sensations presented at the moment, with many ideal sensations which have just actually been experienced, or else are about to be. A visual perception further consists very much in a state of general retinal excitement, and this, no less than the relations of form which are disclosed by it, determines the nature of the state of consciousness into which it enters as an element.

It is a striking fact that most of our visual perceptions are double; either retina receives approximately the same image. This correspondence only holds good for the inner portion of the field of vision; each eye supplements on its own side, right or left as the case may be, the central or common portion of the field. This may easily be seen by closing first one eye and then the other. But further, the difference in the position of the eyes makes a difference in the appearance of solid objects. The right eye sees further along planes facing the right, the left eye further along planes facing the left. The stereoscope depends for its vividness on the combination of photographs taken in differing positions answering to this actual position of the two eyes. Hence, too, plane representations can never exactly produce the impressions which are made on the two eyes by solid objects.

The merging of the images of the two retinas into one visual field, offers the closest combination of sensations into one group that we can find. It is indeed almost paralleled by that of the sensations received by way of the two ears. That either eye or ear contributes its part to the total impression, can be seen by closing the eyes or ears in turn. The combined sensations have a quality different from that of either of the single sets. That two sensations derived from corresponding points in the two retinas should fuse into a single impression, has given rise to many theories in explanation. It has been suggested that the two corresponding sets of impression overlap one another, so to speak, in the mind. But this overlapping is not to be understood literally. Self-observation will show, I think, that the two eyes do not run precisely together; one or the other is always tending to take the lead. The sensations which one eye brings occupy the central, or clearest, portion of consciousness, and the sensations of the other eye fall into the background, and merely supplement those of the former.

78. **Auditory Combinations.**—The organs of hearing, like those of vision, are double. By this means we can distinguish the direction of a sound; it is referred to that side on which we hear most distinctly. By turning the head we find that position in which we hear most clearly, and assume that the sound comes from the quarter to which the head is turned.

The two ears hear together, and similar sensations received by way of them blend in the way described for vision; one ear follows the lead of the other, and supplements the sensations which we receive from it.

As is shown by our hearing chords and, in fact, harmonized music generally, the parts of the ear which correspond to different notes can be simultaneously excited; thus, there is a possibility of combined sensations here, as in the case of organs where the local character is more marked. A reference to § 29 will show that in the timbre of the notes of different instruments there is an approach to an harmonic character. Usually one note in a harmony seems to take the lead, and the others seem to merely supplement it; this is most often the highest note. But practice can enable the ear to follow the lower notes of a harmony.

The ear is especially keen to detect varying intensity, or rhythm, in sound. And I think it will be found that each single beat, in the more common times, has a character of its own. Thus, in three time there arise together with the actual sensation the ideas of the beats lately perceived. That only one in three beats should be emphasized, leaves the other two in the air, so to speak, and it is this consciousness of the long interval between the beats at which one as it were comes to solid ground, which gives the peculiar swing that characterizes the triple time of waltz music, a swing of which we are conscious at each instant, and in hearing each note. In two, or four, time the *ictus* falls more or less strongly on every other note.

Thus auditory sensations enter into combinations in two directions. They are combined in the impressions made upon us by chords, and also in rhythms where they are associated with the ideas of sensations just received. The sureness with which the successions of notes forming melodies can be retained, greatly depends on the arrangement in bars.

79. **Smell and Taste Combinations.**—The scents of a bouquet of flowers afford an instance of a combination of olfactory sensations; such combinations, however, have rarely any intrinsic importance, and so are rarely remembered. There are some cases, nevertheless, which are of interest; thus the combination of odours which marks a druggist's shop, the blended smell of pitch, oakum, machine oil and brine, which haunts a steamer, are characteristic.

Combinations of taste sensations are illustrated by the flavours of various favourite combinations of viands—bacon and beans, lamb and mint sauce, strawberries and cream, beer and cheese, suggest well marked complex impressions.

The example of wine and tea tasters shows that the ideas of past sensations of taste can be brought into comparison with present ones. And ideas of taste and smell can be compared in the absence of the actual impressions. "The scent of lilies I can distinguish from that of violets, though actually, smelling nothing; and I can prefer honey to mead, making use only of remembrance."[1]

80. **Muscular and Tactual Combinations.**—The close connection of muscular with tactual sensations is pointed at in the fact that those parts of the body which are best endowed with organs of touch—the tongue, lips, hands and arms—are also the most mobile. Tactual sensations gain much as sources of information as to the outer world, by being combined with movements; they not only thus give rise to a whole class of composite sensations, but also make possible the knowledge of those relations of the parts of surfaces and solids to one another which constitute form.

[1] *Augustine's Confessions*, book x., c. 8.

One of the most familiar combinations is experienced when a part of the skin, say the tip of a finger, is moved along a surface. Here the muscular sensations derived from the arm are combined with the tactual sensations received by way of the finger. When the surface along which the finger moves is also a part of the skin, say the back of the other hand, there is a second series of tactual sensations, which is received from the portion touched. In this case one set of muscular sensations is combined with two sets of tactual sensations. When the tip of the tongue is pressed against the cheek, there is a similar combination.

81. **Visual and Tactual Combinations.**—The best case of the combination of visual with tactual sensations is perhaps afforded by the association of the appearance of parts of the skin with the tactual sensations which are received when they are touched. Let the reader look at the back of one hand, and touch it with a finger of the other hand. He will then observe the combination of the two kinds of sensation.

Another case is the association of the appearance of external objects with the sensations which contact with them would arouse. The distribution of light and shade on a surface is associated with the characteristic series of tactual sensations it might produce in us, according as it is rough or smooth.

82. **Visual and Muscular Combinations.**—The movement of the yellow point which is requisite for clear vision, is managed by means of the movement of the head as a whole, or of the eyes. It is needless to more than mention the complete turning round of the whole body, or the walking some distance in order to obtain a clearer view. Yet, absurd as it seems to connect the movement of the body as a whole

with the direction of this single point in the retina, namely, the yellow point, to parts of the visual field, it is both appropriate and convenient to treat the matter in this way.

As the body moves along, each change of position is accompanied by a change in the appearance of the visual field. If we are going straight forwards, there is a continual enlargement of the forms which occupy the middle of the field, and as they increase they pass to one side or the other, and then out of sight. If we are moving backwards, as when seated on the back of a gig, the process is inverted; objects come into view from the side of the visual field, and diminish as they approach the centre. This complete inversion of the ordinary experience has the effect of making some people giddy. (Cf. § 28, end.)

Without moving the head, the ocular muscles give the sight a range of about 180°. When the muscles of the neck and shoulders are employed, this is increased to something more than 300°, but still short of the complete circuit. When the trunk is turned round, the legs remaining unmoved, the sight can range over the complete circuit. In all three cases, however, the yellow point fails to obtain a perfect view of the whole field. There is always a portion which is imperfectly seen at the back. And it is only by turning the body completely round that we can scan clearly every point in the circumference.

Now the muscular sensations which are received during these movements are associated with the changing visual sensations, and enter into combination with them, no less than the sensations from the muscles of the eye. Our estimate of angular measurement is based perhaps more on the sensations from the larger muscles, especially of the neck, than on the sensations received by way of those of the eye. As a matter of fact, when the choice is possible the gaze is fixed on an object by turning the head towards it, rather than by moving the ocular muscles, with the head remaining still. And indeed, the sidewards direction of the glance unaccompanied by the movement of the head always gives a sly expression, as may be seen by walking through a picture gallery, and noticing the old portraits, many of which have this fault. It requires practice to restrain the inclination to turn the head round, when intending to take a look at anything slightly out of the direct line of vision. Thus the function of the ocular muscles is limited to a range of a very few degrees in the centre of the visual field, larger movements being effected by the muscles of the neck, the shoulders, the thighs, or by turning the whole body. Now in nearly every case there is a possibility of choice between one movement and another, so that each particular change in the visual field is not connected with one set of muscles exclusively. Thus, my visual field may gain to the left and lose to the right, by the movement of any of these four sets of muscles—the ocular, the cervical, those of the shoulders and thighs, and lastly those concerned in the movement of the whole body.

It is important to note that the movement of the larger muscles is ordinarily accompanied by those of the ocular muscles; the latter take the lead in the movement which then spreads as required to the larger muscles. When I imagine myself to turn my gaze to the left, I am conscious of a feeling of tension in the ocular muscles which colours the whole idea.

The series of muscular sensations accompanying such movements, now in one direction, now in another (§ 75), become gradually associated with the changes in the visual field which such movements bring about. Will the reader be good enough to look straight in front for a moment and to keep his eyes fixed? Let him turn his attention to two points in his field of vision separated by a considerable distance. When he passes in thought from one to the other, he will be dimly conscious of sensations in the muscles of the eye, even if he succeeds at first in repressing the tendency to move the eyes. If he choose the points well towards the outskirts of the visual field, he will probably perceive a tendency to movement in the muscles of his neck as well. In this way the appearance of two points in the field of vision is associated (*a*) with the faint excitement of the surrounding parts of the retina, and especially the intervening part (§§ 76, 77), (*b*) with the muscular sensations accompanying the direction of the yellow spot to one point, after being directed to the other, and *vice versa*. When the distance is small, the ocular muscles are alone concerned. The cervical muscles may come in when one of the points is towards the boundary of the field of vision. The sensations from these more distant muscles are not concerned when the gaze passes from one to another of two points, both of which are in the field, but only when the field of vision passes under a considerable transformation, a large portion passing out of view on one side, and being replaced by a new portion on the other.

The importance of visual sensations in guiding our movements is seen in certain cases of locomotor ataxy; patients are unable to perform many very simple acts if their eyes are closed.

83. **Combinations of Visual, Tactual, and Muscular Sensations.**—We have seen how visual sensations combine with tactual sensations (§ 81) and with muscular sensations

(the last paragraph). It is now necessary to observe how all three classes combine together.

In analyzing the effect of impressions received from the field of vision in the last paragraph, no reference was made to the external causes of such impressions. If however the attention is turned to what the various retinal impressions signify, we shall find it necessary to supplement the enumeration there given. In § 81 it was pointed out that differences of light and shade are connected with tactual sensations.

Thus the excitation of a portion of the retina is associated with (a) the movement of the ocular or other muscles, by which images after passing over the yellow point are brought to bear on the point in question; (b) the tactual sensations which we should receive on going up to and handling the object (these are especially connected with the light and shade of the object); (c) the muscular sensations we should receive on walking from our present position to the object. These last are connected with the relative magnitude of the retinal image, objects of which the images are smaller being further off, and so demanding a longer series of movements. The clearness of the image and its brightness, the extent to which it is covered by the images of other objects, are all signs by which we infer the amount of walking or other movement necessary to reach the object.

From these considerations it becomes evident why sight is the chief channel of external perception. The organ of sight responds to changes in surrounding objects, even those that are very slight, or very distant. At the same time the full meaning of those changes is not disclosed by the changes in the coloured surface of the field of vision.

Retinal impressions are important to us, as symbolizing the movements we must execute to pass from one object to another, or the impressions of contact which surrounding objects are likely to make upon us when we reach them.

The combinations formed with other kinds of sensations by those of taste and smell, are illustrated by the combinations of impressions which constitute our perception of external objects, and need not be specially referred to.

84. **Auditory Combinations.**—Sensations of hearing do not so readily combine with other classes of sensations. The chief instances of their combination with visual sensations are found perhaps when we see two bodies collide and hear the report. Then there are the cases of musical instruments, of the human lips, arousing the ideas of musical sounds or utterances. But these connections have nothing of that intimate character which distinguishes the combinations treated of in the last paragraph.

There is however a large class of combinations into which auditory sensations enter, which are exceedingly intimate. Thus the utterance of a word involves (*a*) muscular sensations from the vocal chords, (*b*) the hearing of its sound; these may be further associated with (*c*) its appearance in print or writing (visual sensations), and (*d*) the muscular sensations accompanying the act of writing it; (*a*) and (*b*) usually suggest one another and are generally suggested by (*c*) and (*d*). The closeness of the association of the appearance of a word with its utterance is shown by the difficulty felt by individuals who have not had much practice in reading, in following the words in print without muttering. On the other hand, the pronunciation or sound of a word does not of necessity bring up the movements necessary to write it, or the idea of its appearance in print.

85. Perception of Surrounding Objects.—The various objects in our environment affect us now through one sense, now through another; now for instance by the eye, now by the ear. In this way the impressions we receive become associated together. Those combinations we have just been considering depend on the presence and stimulus of external objects. Continual intercourse with our surroundings brings us from time to time under the influence of the same stimuli, causing in us particular combinations of sensation. These combinations stand in our minds as representations of the answering external objects; in other words external objects are for the psychologist "permanent possibilities of sensation," to quote Mill's well-known phrase.

Take my clock as a convenient example (the invitation is metaphorical of course); its colour and form are associated in my mind with sensations of contact and resistance, and both these again with the noise of its ticking. Now whenever I receive any of these impressions from it, be it the sound of its ticking, its colour, or its hardness, which affects me, those sensations which I do not immediately feel, are faintly recalled along with the actual sensation of the moment. Thus there is combined in the percept of the clock, sensations and ideas of sensations. At this moment I perceive the clock ticking; this is an actual sensation. At the same time I have the ideas of colour and resistance, which the sight and touch of the clock would actually give me. Looking up at it, the visual ideas become actual sensations like those of which they are the revival. Putting my fingers in my ears and looking up at the clock, the old relation is inverted, the auditory sensations are now merely ideal, while the visual ideas are become actual sensations. When I go out of the room, all the possible impressions which the clock might give me are now ideal so far as I think of it at all. Thus there arises in my mind a group of ideas answering to the clock. When, on returning and hearing the ticking, I perceive the clock, what really takes place is this—the impression of the ticking recalls the ideas of the impressions which have previously accompanied it. It is this reference of an impression or idea to the group of which it forms a part, that constitutes perception.

This process might be illustrated by any other kind of object—horse, dog, tree, &c. Observe that we recall not only the groups of impressions which any particular object may have given us, but also those which other objects of the same kind may have given. The group of ideas which fills out our perception of a tree is a representative image of many trees. The way in which similar groups of impressions blend into a single group, which represents and typifies them all, has been explained in § 48.

Not only, however, do single objects present us with groups of impressions and ideas; we also find that they affect us one after another in certain ways, and thus we obtain a general notion how they stand to one another. We find, for example, that to reach and handle some objects we must come into contact with, or pass by, other objects. Thus in going from my table to the clock, I may reach it by passing a bookcase; or taking a longer détour, I may arrive by way of the hearthrug and round by the door. When, then, I say that I perceive the clock to be in such or such a position, what is meant is this—there is a certain order in my impressions, such that I receive sensations of contact by touching the clock after coming into contact with other objects. In other words, the order in which external objects impress the mind depends on the way in which they are actually grouped, and the order of our impressions is an index to this grouping.

When therefore we speak of distance and direction, of number and magnitude, as being perceived, we are using abstract expressions; what we actually perceive is, that some objects are near or far away, are on this side or that, are great or small, are numerous or few. In the *Iliad* we are told of cities with broad streets, of kings with broad realms, of the broad gates of the House of Death, but the word "breadth" does not occur; nor do such abstract ideas present themselves to the mind except under the stimulus of some special study.

86. **Distance.**—We become conscious of distance through the number and kind of the movements necessary to pass from one object to another, and these of course are made known to us by muscular sensations. The perception of short distances can be gained by the movement of a single limb—arm, hand, finger, for example. We often judge of the distance between the parts of some object by passing the hand along it. Here touch sensations supplement the muscular ones in determining our estimate of distance.

Usually we take our sight impressions as a guide and check them by reference to muscular sensations when necessary. The distances between the images of objects on the retina, and the changes undergone by those images as our distance from the objects varies, symbolize the movements we must make in order to reach them.

87. **Direction.**—We have a standard by which to measure direction, in the angle through which our eyes must be turned, in order to bring an object the direction of which is in question, into our direct line of sight. This is the

PERCEPTION.

way in which we measure small divergencies from a direction straight in front of us. We can also measure direction by the extent to which we must turn the whole body as explained in § 82. In this way we become conscious of greater divergencies from our line of sight.

88. **Continuous Substance: Size.**—If the hand is moved along the surface of an object, the sensations of movement are accompanied by sensations of contact. The larger the surface along which the hand is moved, the longer the sensation of contact. Thus our sensation at any moment is supplemented by the ideas of the sensations just received, and as this ideal element increases, so we estimate the size of the object which has given rise to the sensations. These continuous touch sensations answer to retinal impressions. According as an object can affect us with long series of touch impressions, so do we find that its image covers a large portion of the retina. When objects are very large, it is necessary to execute long series of movements before we can become conscious of their whole surface. The number of paces we must take to survey a large object in this manner measures its size.

89. **Discrete Substance: Number.**—It is very rare that a single object covers the whole of the retinal field; usually we receive impressions from many objects simultaneously. Their separateness or *discreteness* is indicated by a change in the character of the impression, a difference in quality or depth of colour, or the intervention of part of the background. Similarly, if the hand is laid on the surface of an object and we walk along it, a few paces will generally suffice to produce a break in the continuous touch sensation. The hand will begin to move freely through the air, and perhaps after this break receive a new series of touch

impressions. What we call number is made known to us by these breaks in our sensations.

Simple as the idea of number seems to us, it involves a certain power of abstracting the attention from the other qualities of objects, and is beyond the capacity of minds of low development. In practice the Damaras "make use of no number greater than three. When they wish to express four, they take to their fingers, which are to them as formidable instruments of calculation as a sliding rule to an English schoolboy."[1]

90. **Solidity.**—Our perception of solidity may be analyzed into a series of sensations of movement followed by a sensation of resistance; or a sensation of resistance may be associated with the ideas of the onward movements which the resisting body prevents. The reader may easily observe this by moving his hand until it comes into contact with some solid object.

91. **Force: Power.**—Our perceptions of force and power take their rise in the consciousness we have of our motor impulses. By putting them forth we produce changes in our own position in relation to other objects, or in their distribution, or in our perceptions of them. And then finding that external objects can affect one another or ourselves, in the same way as we on putting forth effort can affect them, we associate the changes thus produced with the exertion of similar efforts (motor impulses) in the bodies which produce those changes.

92. **Change.**—We become conscious of change by reason of the persistence of our impressions in the form of ideas. Side by side with each succeeding impression is the idea of those just past. If there is any difference between them,

[1] Galton, *Travels in South Africa,* c. v.

we are conscious of a contrast between our immediate impression and the ideas that accompany it. And this contrast presents itself to us as an impression of relation. We have such impressions of relation on passing from one impression to another; we have similar impressions of relation on passing from an idea to an impression, or from one idea to another. In this way we become conscious of likeness and unlikeness. When we receive but a slight impression of change or none at all, on passing from one impression or idea to another, we call them like or the same; when we receive a greater impression of change we call them unlike.

Changes in our surroundings make themselves known to us through the fact that, on repeating the movements which formerly subjected us to particular external influences, those influences no longer affect us at all, or if they do affect us, then in different manner and amounts. In this way we distinguish external causes of changed perceptions, from those which depend on us; we may lose sight of an object by its being removed, as well as by closing our eyes.

93. **Succession.**—The perception of succession is implied in that of change. Each new impression which was felt side by side with the idea of the preceding one, becomes in turn an ideal element besides a new actual impression. In this way, each actual impression is accompanied by a number of ideal ones of varying degrees of vividness, the later ideas being the more vivid.

Past impressions are roughly referred to their time of occurrence, by representing to ourselves the amount of impressions we have received since they were freshly presented to us. At the end of the fourth chapter it was pointed out how, in estimating the lapse of time, we are governed by the nature of our impressions. In the absence of counteracting causes, we refer those impressions to a more distant past which are overlaid with the greatest mass of ideas; this is the unconscious method. In consciously determining the time of some experience we generally make use of some striking event in our history as a landmark, and follow the trains of ideas leading up to, or starting from, it.

Since our trains of memories unfold themselves in succession, and generally in correspondence with the course of the original impressions, we are constantly perceiving ideas to have their places in series, and refer the actual events to which they correspond, to their places in a series which is independent of us.

94. **Space and Time.**—The possibility of movement now in one direction, now in another, is represented to the developed mind by a vague sense of freedom, of scope for movement. We have already seen how the several kinds of retinal impression are associated with the corresponding kinds of movement. In this way the visual field becomes a summary, so to speak, of trains of movements, and takes a leading part in our idea of space in general. Our idea of space, then, is not so abstract as is sometimes thought; it is complex indeed, and gained from touch, the muscular sense, and hearing, as well as sight; but its characteristic quality is given by so simple an element as the mere sense of freedom to move our bodies. Add to this, that in moving about we become conscious of an order in our impressions which does not depend on us.

Our idea of time depends upon the fact that experiences are rarely, if ever, repeated exactly. Hence, when sometimes we are performing actions that we have often done before, we are confused by the likeness between the present and some previous experience, and are doubtful for an instant at what point of time we are. It is the constant change in our surroundings which makes it almost impossible ever thus to repeat the same experience. Our states are always changing in obedience to the laws of mind; but the effects so produced are constantly being interrupted and traversed by stimuli and influences beyond our control. These external influences, too, are always changing, and the order of their changes is mirrored in the order of our experiences, and so the manner in which our memories follow one another stands for changes, not only in our own life, but in the course of external events. The root of the idea of time lies then in the constant change of our impressions.

95. **Standards of Sense Perception.**—Scientific observation stands to our ordinary manner of perceiving somewhat as the syllogism stands to our ordinary manner of inferring; it is the type, or standard, to which we should conform—not the picture of an actual process of thought.

The first requisite is that we should accurately refer each impression to its cause; this is the same as to refer it to the answering group of ideas by which we conceive its cause. An eclipse must suggest to us the idea of an opaque body intercepting the light from a luminous body. If it suggests, as to a Chinaman, the idea of a dragon devouring the sun, the reference is unscientific.

Our perceptions will further assume a scientific character in proportion to their completeness or extent, their thoroughness or minuteness, and the rapidity with which they are performed.

The *extent*, or completeness, of a perception may be measured by the number of impressions grouped together in a single operation. Its *thoroughness* may be measured by the exactness with which each element in the impression is observed. Some persons can grasp an object in its general outlines, and fail to apprehend the details; others, again, have an eye for details, yet fail to group them together. These two qualities might be tested by setting the person experimented on to draw from memory an object which he had seen but once. The *rapidity* with which we perceive is an excellent index as to presence of mind. But this quality does not necessarily accompany the previous two. Says Rousseau: "I can perceive nothing of what I see; I can only perceive what I remember, and am acute only in my memories. But gradually all returns to me, and I recall the place, the time, the accent, the look, the gesture, the circumstance; nothing escapes me. Thus, from what has been said or done, I discover what has been intended."[1]

The acts of perception, which consists in bringing impressions, or ideas of impression, together, and comparing them as to their amount or intensity, can be made the object of experiment. The perception of space-magnitudes, for example, may be tested by giving the person tested a rule provided with a moveable pointer, and instructing him to put the pointer exactly half-way, so as to bisect the rule. We may test perception of time-magnitudes by striking the table at intervals of ten seconds, for example, and calling upon the subject of the experiment to do the same without referring to his watch. Here the percept of the given interval has to be retained as a guide in measuring off a similar interval. Further tests may be devised by setting the subject of them to mark off definite

[1] *Confessions*, Book iii.

proportions, say one-half or one-third, of given times. The comparison of colours, weights, textures, may be experimented on in a similar way.

96. **Perception of Ideas : Reflection.**—Hitherto we have been considering states of mind in which the vivid nucleus consisted in sense impressions, round which associated ideas were grouped. We now turn to states in which the attention is no longer occupied with sense-impressions, but is fixed on some idea or ideas.

To reflect is to hold some idea, or group of ideas, very clearly in view, while associated ideas gather round it. What is called connected thought is simply this maintenance of some train of ideas before the mind. Inability to think is inability to keep the attention from wandering from one group of ideas to another. In reflection, then, the vivid nucleus consists in ideas.

We can distinguish between the vivid and the faint revival of past impressions; the same train of memories may on one occasion be reflected upon, that is, clearly perceived; on another occasion we may be barely conscious of its recurrence. In this way the events of our own mental life become the objects of memory; we can recall the occasions on which we have dwelt upon some train of ideas, no less than the occasion when we received the impressions on which it is based.

Clear and distinct ideas are their own evidence. We do not need to know that we know. To ticket with arbitrary names the various states of mind may be useful in the study of mind; it is of little use in the operations of mind. To tell any one that he is perceiving, or thinking, is merely to associate an abstract name with the particular instances of it, and has often the effect of encumbering the free movement of thought. The study of mind can help but indirectly in the conduct of the understanding.

97. Introspection.—We may now supplement § 72 in its references to Internal Perception or Introspection. To look within is to withdraw ourselves from contemplating those external objects which affect the senses. We no longer refer each impression to the ideas of associated impressions. We also refrain from following out the trains of ideas to which each present idea might give rise. By the limitation of the mind's activity in these directions, we can extend it in others. We become clearly conscious of all the impressions and ideas which are immediately present to us, including impressions, such as the common sensations, which ordinarily do not occupy our thoughts, except very faintly. In this way we obtain a group of elements which, if the act of introspection is very thorough, are grasped or perceived in a single "general impression."

Judgments on our fellows, works of art, public policy, are often based on such general impressions of our state when under their influence. The reader will easily distinguish between judgments so formed, and those attained by reflection on the conditions which give rise to these general impressions. We unthinkingly praise or condemn under the guidance of our passing mood, like the judge who always hanged prisoners tried before luncheon. Reflection will indicate to us what part of our general impression may be taken into account, what passed over; we shall allow for bias, ill-health, and, if we are very honest with ourselves, for the limitations of our knowledge and experience. The supreme act of honesty, and the hardest, is to remember that however carefully we have formed these judgments, they are personal, and taken from a single point of view.

CHAPTER VIII.

THE FEELINGS.

98. **Simple Feelings; their Nature.**—As we have followed the path of intellectual development, from its starting point in the sensations to its goal in the most complex and far-reaching reasonings of which we are capable, our regard has been fixed on sensation and thought alike, as informing us about our surroundings and ourselves; we have never inquired how far such states enter into and constitute our well-being or misery. That inquiry we are now about to pursue.

In passing, let us note that mental states are called feelings so far as they have this interest for ourselves. The more complex states of feeling are sometimes marked off as emotions and sentiments.

Sensations differ from one another, not only in quality, as a sound from a colour, and intensity, as a loud sound from a soft one; they are further distinguished by their tone of feeling or interest (§ 21). The sense of sight is a source of one kind of interest, the sense of smell of another kind. Some classes of sensation are in themselves more interesting and more attractive than others. The pleasures of the palate for a person who is hungry, are certainly more engrossing for the moment than the rarest combination of sweet sounds. This attractiveness depends on circumstances, and is not the same for all persons. Some have "no ear for music"; others have no eye for pictures.

Painful sensations present a negative kind of interest; they repel us as much as other sensations attract. This repulsion, too, differs in different circumstances and for different individuals.

Now we may imagine a sensation attractive at first becoming less and less attractive, until it becomes positively repellent. At some point in its downward path, it will pass from one state to the opposite; from being attractive it will become unattractive, and later on positively repellent. The question whether sensations of a neutral kind, neither attractive or repellent, are to be considered as possessing interest—that is, importance for our happiness—has been much discussed. Directly indeed they do not—they are assumed to have no power over us, indirectly they do possess importance, so far as we are conscious of them. For they enter into states with other perhaps more interesting elements, and so determine the intensity of the whole state. It makes a vast difference whether a pleasure or pain is experienced alone, or in company with indifferent impressions. Pains and pleasures are generally diluted; we rarely get them unmixed.

States of feeling, then, are composed of elements of varying degrees of interest; and while the more vivid or interesting portions dominate us, we are also impressed, if less strongly, with their accompaniments. The analogy of this to the facts described in the case of attention will not escape the careful reader, for indeed we are merely considering the same mental states under two different aspects. The elements which enter into emotional and intellectual states are the same, but while clearness and distinctness is the leading characteristic of intellectual states, interest is the leading characteristic of emotional states.

The highest intellectual operations depend on a certain balance of the tendencies at work in the mind, a balance which is endangered when any single tendency or group of tendencies unduly dominates us. On the other hand, emotion is most evident when we are so dominated. It thus implies a disturbance of the mind's accustomed state, a "movement out" of it. A state of mind, then, is emotional, so far as any single tendency, or group of tendencies, prevails to the exclusion of others. Pleasure and pain are regarded as specially characteristic of emotion, because association with them usually establishes such undue prevalence.

99. **Physiology and Expression of Emotion.**—Mental disturbance implies nervous disturbance. Emotions are usually connected with more or less violent nervous excitations, which manifest themselves in different ways—sometimes in changes in the physiological processes of circulation, respiration, and secretion; sometimes in motor impulses.

And first, as to the physiological processes. Certain emotions are associated with certain organs of the body. "Pain and grief affect the lachrymal glands; anger, the liver; fear, the stomach; uneasy expectation, the renal and connected passages."[1] Add to these examples, the acceleration of the heart's action, which marks some emotions, and its retardation, which marks others; the variations in the rate of breathing, which accompany fear and other forms of excitement; the relaxation of the muscles controlling the arteries, which causes blushing, and their contraction, which causes pallor; also the increased activity of the sweat glands—

[1] *Wundt*, vol. ii., 407.

and we have a few of the ways in which the rest of the organism answers to the disturbed state of the nervous system. The functions enumerated are not usually under the control of the will; and their connection with nervous states is a roundabout one, produced through the nervous centres which control these functions, centres probably situate in the medulla oblongata. Such reflex activities, following upon nervous excitations, are less obvious, and therefore less easy to observe, than the muscular activities next to be considered; but their very involuntariness renders them valuable sources of information when the person observed cannot be relied upon to be candid. Lombroso, the famous Italian authority on crime, has tested the sensibility of criminals to various thoughts and emotions with the plethysmograph. This is a delicate instrument for measuring mental excitement, and depends for its use on the fact that the slightest emotion causes an alteration in the amount of blood present in any part of the body. By noting the effect of various suggestions, he arrived at the conclusion, that the strongest impressions (superior to the normal) were produced by cowardice, fear of the judge, favourite modes of indulgence, but above all by vanity.[1]

Secondly, as to the muscular accompaniments of emotion. In the normal state of the nervous system, the lower centres, which control the several muscles, are themselves held in check by the higher centres; otherwise, each sensory impression would produce an answering motor impulse. Moderate impressions, however, do not usually give rise to any such movements. But as the nervous system is more and more disturbed, the muscles begin to be set in motion, and this motion calls forth an increased effort to

[1] Havelock Ellis, *The Criminal*, p. 122.

THE FEELINGS.

check it, and thus its course is marked by alternate control and want of control. The movement spreads, as the excitement increases, from the smaller and more easily moved muscles to the larger ones. The order and extent to which these movements are combined differs for the several emotions, and for different individuals. Mr. Darwin, in his well-known book, has pointed out how the different facial muscles combine to produce the different forms of facial expression. Not only facial movements, however, but all movements, so far as they indicate to other persons our mental states, are forms of emotional expression.

The reason why particular feelings should be expressed in particular ways, lies partly in the structure of the nervous system, by which certain disturbances in some parts call forth certain disturbances in other parts. It lies also in the effort to gratify the emotion or to relieve it, which gives rise to what we may call the natural language of emotion. Instances of this may be seen in the movement back which we execute on receiving some shock; the gesture of striking employed by an angry man; the raising of the arm as if to ward off an imaginary blow, by a person who is alarmed. The effort to hide the signs of emotion forms itself a secondary emotional language; the compressed lips and the clasped hands are themselves signs of the emotions whose natural expression they prevent. Lastly, the imitation of the gestures of others further extends the range of our emotional language. Some forms of gesture, such as shrugging the shoulders, in this way spread from the persons by whom they were first employed. Imitation also helps in the development of the natural forms of expression. Whatever may have been the original connection of language with emotional expression, its use is acquired at this present stage of development entirely by imitation; the child must *learn* the use of I and Me.

The expression of emotion sometimes deepens it; we can often make the emotion real of which we simulate the expression. On the other hand, by repressing the signs of emotion, anger for example, we are sometimes enabled to dispel it.

Emotion is sometimes relieved by being given utterance. A half-idiotic youth in the Boston lunatic asylum, U.S.A., was the subject of violent and frequent paroxysms of anger, and, with a view of moderating these, it was suggested that he should be kept for some time every day in rather fatiguing exercise. Accordingly he was employed for two or three hours daily in sawing wood, to which task he made no objection. The paroxysms never displayed themselves except on Sunday when his employment was intermitted. As it was considered, however, to be better for him to spend a part of that day in sawing wood than to be irascible during the whole of it, his occupation was continued during the whole week, whereupon he became completely tamed down and never gave any more trouble by his passionate displays.

100. **Complex Feelings; their Classification.**—States of feeling arise in the same way as states of thought; that is to say, by the elaboration of mental elements into more or less complex combinations. And we find in consequence that we can distinguish combinations of feeling in the same way as combinations of ideas. We can experience trains of feeling no less than trains of thought. Then again, feelings combine into clusters. It follows that the laws and kinds of association which have been described in connection with intellectual states also hold good here.

THE FEELINGS.

The one characteristic feature in emotion is what we have called its power over us, its interest. Hence in distinguishing emotional states from intellectual ones, we shall be chiefly concerned in tracing how far the interest of a mental state interferes with its intellectual character.

The classification of the feelings is a subject on which there is no general agreement. Mr. Bain is content to "trace out the most usual forms and varieties" of emotion, and enumerate them side by side, like the species of a genus of animals or plants; he therefore calls his method the "natural history" method. Mr. Spencer proposes to classify emotions, according to the number of stages by which they are removed from the sensations in whose revival at first, second, or third, hand they consist. This distance he denotes by the name "remoteness from sensation." Thus the emotion of terror involves the representation to ourselves of some hurt as likely to befall us, and is therefore more representative than a feeling like the sensation of a sweet smell, which need not involve anything but the immediate impression. On the other hand, the emotion of terror is usually connected with the idea of some particular hurt, as affecting us, and is therefore less representative than a sentiment like the love of justice, into which the ideas of single sensations no longer enter. Such feelings are highly representative; they involve, not the reproduction of single sensations, but of feelings and thoughts, which consist in the *ideas* of sensations, and not in the sensations themselves.

The classification followed in this chapter is a combination of the two kinds described. We enumerate the various elements of feeling, and follow them out into their various combinations as they become more remote from sensations.

Feelings, as we said, derive their general character from being very intense. Now this intensity may be caused in two ways, and to these there answer differences in the quality of the feelings.

On the one hand, we shall have the emotional states to which the mind is not already adjusted, such as surprise, wonder, admiration, veneration. In these the mind experiences a more than usual shock when passing from its previous state into a new experience. Such feelings take their rise in impressions of relation (§ 31). At their first rising, we are conscious of a shock, but not of their nature, pleasant or painful; as we recover our balance, this nature is disclosed. Sometimes indeed the contrast between two uninteresting states may give rise to surprise in which pleasure or pain does not enter. The average schoolboy is often surprised at the solution of some problem he has failed to do; but he is not interested. So far as these feelings of surprise are coloured by pleasure or pain, the memory of them may cause emotions of the kind about to be described. Obviously, however, they are well-nigh incapable of reproduction; their force depends for its very overpoweringness upon their being presented to us for the first time; and therefore they are transient. Only when the mind remains open to the contrast of unselfish and heroic conduct with average performances, can it continue to admire and to venerate. The contrast becomes dim with use, and blends, for dull eyes, into the monotony of the commonplace.

Opposed to the feelings of surprise stand those feelings which arise when the mind is preadjusted to its experiences—the feelings of memory and expectation, such as hope, courage, pride, contempt, anxiety, fear, and despair. These depend on the revival of past trains of feeling, to which some present suggestion gives rise. The mind projects its past tones of feeling into the future. In order that these moods may be thus present from time to time, dominating our view of the future, they must have some intrinsic attractiveness, some close connection with our happiness, or some root in habit.

Leaving out of account those emotions which depend on chance fixity of idea, and are therefore morbid, we shall find that the intensity of emotion is determined in the long run by association with *our own* pleasures and pains. At first, all tone of feeling is personal and selfish. Out of this there grows a more complex state, in which we take account also of others' feelings. In becoming more intellectual, this state becomes also less intense. It is only growing imagination that can enable us to have truly deep as well as wide sympathies; our thoughts of other persons only become truly sympathetic when we represent ourselves as doing or suffering what they do or suffer. Complex as they ultimately become, these feelings take their rise in the simple pleasures and pains enumerated in Chapter III. And a reference back to these simpler experiences is needed from time to time, to make our complex states as vivid as possible.

In either of these main classes—feelings of surprise, or of intrinsic interest—we can use Mr. Spencer's standard, and range the various species according to their representativeness; thus the growing representativeness of the latter class of feelings is manifested in the broadening of our sympathies; the feelings of surprise rise into wonder, awe, and veneration as they become more complex and representative.

Sometimes all lines of division are confused, and single states present several or all of the different characters we have just considered. These marks should thus be regarded as points of view convenient for comprehending emotion, and not as exhausting all the possible species. We can have as many feelings as we have thoughts, and it is no more possible to make an exhaustive description of emotions than it is to lay down limits of thought.

101. **Trains of Feeling.**—Trains of feeling leave their traces in the same way as chains of ideas. When we have experienced a succession of pleasures and pains, we tend to go through the same series again, a suitable suggestion being presented. The more familiar persons and objects in our surroundings are constantly affecting us in various ways, and consequently the sight or thought of them becomes the starting point of many trains of feeling. These trains of feeling converge, diverge, and intersect in the same way as trains of a more intellectual nature (see § 51). They thus tend to pass into one another and to blend, to conflict with one another by drawing the mind different ways, or to run side by side. And we might conceive the mind to go on elaborating emotional states when it was no longer directly affected by pleasure or pain. The course of our emotions is thus determined from within as well as from without.

Now trains of feeling consist of the same sensations and ideas as those which compose our trains of thought, only that these sensations and ideas are so strong as to take on a pleasant or painful character. What will happen when these more intense series begin to accumulate round given starting points of suggestion? Like the more distinct intellectual trains, they will tend to coalesce into representative series—series which do not represent any single past experience of the respective kind, but in a manner stand for all such series. There are formed in our minds systems of feelings in the same manner as systems of ideas. We thus meet new experiences with varying tempers, in the same way as we bring preconceived ideas to bear on them; and the temper is simply a preconceived idea with emotional colouring. In this sense concepts are also moods.

102. **Emotional Suggestion.**—At first the child's mind is in a continual state of wonder. Experiences as they come to it call up no answering ideas. It has yet to see how the new experience stands to the old ones, and how it will affect its happiness. The case of the adult is very different. He is provided alike with preconceptions and moods for almost every fresh occurrence. "Without resource he meets nothing that threatens him." Each suggestion sets his memory to work along paths where every step is rich with emotional force of a subtle kind.

Of course there are great variations. Some moments take us dull and irresponsive; and some minds are in that state of callousness so abhorrent to Greek susceptibility. At the other extreme, the mind is too easily moved from its balance, its purely intellectual temper.

Emotional activity implies the presence of a large mass of associations to which each new feeling may attach itself. We have to meet suggestions half-way. We can only receive so far as we ourselves contribute. And to match this, we can always find what we look for; life will be depressing, impure, commonplace, according to our temper; or inspiring, high, and pathetic.

103. **Emotional Conflict.**—The very state of emotion implies that some tendency or group of tendencies is in possession of the mind to the exclusion of others, which in more balanced states of mind would have been present to us. And these last produce a state of mental tension the

stronger in proportion as the excluded tendencies are the stronger. Thus if we engage in some course of action, which is at variance with our ordinary inclinations or ideals, the thwarted inclination and the forsaken ideal show themselves not indeed in full and distinct forms, but in a vague sense of uneasiness; this may be so slight as to pass away, or so strong as to impel us to seek some release from it. We may gratify the inclination, or return to our old ideals, hardly so much for their own sakes as to end a state of mind which has become unendurable.

104. **Harmony of Feeling.**—When, however, there is so great a preponderance, on the side of one set of feelings, that contrary feelings can find no room in the mind, there arises a mass of feeling which is scarcely an emotion in the sense of being a mental disturbance. In such a mass, all the mental tendencies which under ordinary conditions would have been objects of thought find realization, and thus the operations of thought can proceed in their normal manner.

We experience states of this kind, when our daily occupations and our daily thoughts are busied about tasks in which inclination, self-interest, and duty all point the same way. These states are rare, however, and we are usually in a state more like the conflict described above.

105. **Reflected Emotion.**—When we are in such a state of harmonious feeling, the central emotional colouring radiates upon, and is reflected by, outside ideas which stand in no necessary connection with it. The aversions or preferences of which sometimes we are the surprised possessors come in this way. The flashes of happiness which leave their marks in our lives, make every circumstance connected with them interesting with a transferred attractiveness.

106. Moods and Temperament.—The undercurrent of pleasure and pain which flows beneath the general stream of our thoughts, does not change its direction so rapidly as they; its rhythms are longer. Hence we often regard some new experience with the feelings aroused by the preceding quite different experiences. Thus there arises a want of correspondence between our moods and our surroundings. In later years we view life with the temper begotten of experiences a generation old.

The persistence in a train of feeling on many successive occasions may give it such power as to derange the mind. Indulgence in violent fits of anger produces in women a form of insanity which is well recognized by physicians.

At first the feelings change from one kind to another with considerable freedom. "The young," says Aristotle in his *Rhetoric*, "desire passionately but quickly cease from their desire." Yet even in the earliest years certain moods are becoming habitual, which as they gather strength will constitute temperament. According as we are habitually hopeful or despondent in early years, habitually self-controlled or uncontrolled, so will our temperament be; we shall be sanguine or melancholy—resolute or wavering.

107. Feeling and Thought.—Feeling is not the necessary enemy of thought as is sometimes suggested. It is only when our feelings are in conflict that the orderly activity of the mind is hindered. Feeling is so far from being unfavourable to thinking, that it is necessary in order that we may make any sustained effort of thought. The divine afflatus of the poet, the sense of vocation felt by great leaders, are the shapes taken by the feelings which give being to their works. Only let the feelings set in one full harmonious stream towards some goal, and they will sweep the thoughts very far to its attainment.

Turning the statement round we may say that no great purpose, intellectual or practical, is ever carried into execution except under the influence of feeling. There must be the hours of insight when the mind's glow illumines itself and its object. Our ideas must rise so clear to view that they dominate the attention, and so strong that they hold their own against conflicting ideas.

108. **Selfish feeling.**—At first we can only have strong feelings in proportion as our own experience is touched. Pleasures and pains are pictured in the circumstances in which we have had them, and therefore in connection with ourselves. Children are delightfully incapable of entering into the feelings of their elders. Mr. R. L. Stevenson describes somewhere the callous way in which a child came up to him as he lay ill, and airily disregarding the sufferings which he was engaged in contemplating, requested his attention for some more pressing business of its own. And this selfishness is but veneered over, even in later years. Even when we are most sympathetic we do not really step out of ourselves. The implicit thought which runs, even through each less selfish emotion, is that of our own self as being hurt or benefited, feared or despised; or we think of ourselves as causing the emotion by our own acts, as the injurer or the benefactor, the awarder of praise or blame, or the person fearing or despising. If we wish to excite ourselves to the highest possible sympathy, we try to put ourselves in the place of another; and in proportion as we transfer our feelings to his imagined case, so is the strength of our sympathy.

Thus it appears that we must have undergone the feelings

ourselves with which we would most perfectly sympathize. The mental excitement involved in such circumstances as these leads on, by an inherent necessity, to some expression or other. For the typical mental operation is neither thought, feeling, nor act taken singly; it is an impression followed by a movement (§ 13, end). Hence when we hear some moving tale, the faint emotion which stirs in us leads on to the answering acts; indignation, for example, is often indicated by the clenching of the fists. In cases like these, our emotion having no occasion of our own to be spent upon, is relieved by being spent in the way suggested to us. It is no subtle calculation, therefore, of our own and our neighbours' interests which determines us to show our sympathy in act; the mere intellectual presentation of ourselves as being in another person's case, tends to make us act for him. Experience, of course, soon schools us to restrain our sympathetic inclinations; we lose the habit of 'gushing.'

The first beginnings of the life of feeling consist in the gradual extension of our own feelings to new objects and to new cases. Surely, therefore, it is somewhat paradoxical to use the complex concepts of self-interest and property, only gained at a later age, in order to explain them. That there should so often be a conflict between our own feelings and those which would be dictated by another person's circumstances, is an important factor in the development of feeling, but it is not the first to come into operation. The limitation of children's feelings to their own case is an intellectual limit; their minds are not yet expanded to the larger expediencies. Theirs is a natural and excusable selfishness. Selfishness, in a bad sense, arises when with a wider intellectual horizon our feeling remains cramped in that corner which more concerns ourselves.

O

109. Emotional Interpretation; Tact.—The life of feeling does not consist, any more than the life of thought, in the mere repetition of our experiences in their original order. Under ordinary circumstances there are two main occasions on which we build up our own experiences of feeling into new combinations; firstly, in interpreting the signs, intended or involuntary, employed by the persons about us, secondly, in picturing the emotional life of imaginary characters or persons we have never seen. Language, of course, is the usual vehicle by which we communicate our states to one another. And by its means the same, or similar, intellectual objects can be brought before many minds. But there is one thing which language cannot do; it cannot convey the general impression which each group of circumstances makes upon us; it exhibits our feelings in outline without regard to their perspective. On the other hand, the general emotional expression is an index to the state of the mind as a whole; and with it for guide we catch the broader groupings of another's mental landscape. For it is not enough to have had the experience, whose effects we infer to be in the mind of another; we must also know the proportion those effects bear to the traces left by other experiences. Things that loom large from one point of view look trifling from another. This appreciation of the exact relative importance of various feelings is *tact*.

To interpret rightly the emotional expression and the words of others, we must keenly observe them in the first place; in the second, we must have had similar feelings to those of which they are the sign; and lastly, we must attach the same meaning to the same forms of expression. Thus observation, inference (from the signs to the feelings), and emotional construction are all required.

Similarly it is no simple matter to understand aright poetry, novels and history, which portray the feelings of other persons, real or imaginary; it requires wide sympathies and a powerful imagination. It is absurd, then, to expect to appreciate such works without due preparation.

110. **Sympathy.**—The emotions which at the root consist in ideas of ourselves as receiving impressions, or as putting forth activity, gain very much in complexity, when by the process of interpretation just described they are built up into the ideas of other persons as receiving impressions or as exercising activity. We may mark off three kinds of emotion in proportion as the idea of other persons enters into it.

The selfish emotions are exclusively directed to our own interests, and take no account of the feelings of others. Such, for instance, are the love of power, or of possession.

Next comes a class of emotions in which regard for our own and others' interests is combined; these take account of others' feelings as a means to their own end. Such are the love of praise, the resentment of unprovoked injury. Praise is sought for the rewards it often brings. Unprovoked injury is resented because it may fall on us in turn.

Lastly, we have the disinterested emotions, in which the conscious reference to self has well-nigh vanished: the sentiment is not limited to the cases in which particular individuals are conceived to suffer the pain, or to enjoy the happiness. The idea of any one enjoying a pleasure is now itself a pleasure; the idea of any one in pain is itself a pain.

111. **Sentiments.**—But the mind does not only feel emotion when the competition between its own and another's interests is in question. All the ways in which it realizes

itself may receive emotional colouring and be qualified by the idea of other minds as similarly engaged. The ideas of ourselves as resolving, and acting as critics and judges, as citizens, and so forth, are extended by taking account of others as exercising similar activities. But observe that here, as in the case of the pure concept, the leading one among the convergent associations is taken as typical and representative of the rest; the idea of one's self being so much stronger than that of other people necessarily takes the lead.

The sentiments may be defined as those feelings which answer to the most abstract ideas—that is, to the most abstract concepts of ourselves and others as doing or suffering. (Observe that this personal reference is essential. We do not ordinarily entertain sentiments with regard to pure abstractions, or things which do not concern us.)

The sentiments, like the less complex feelings, arise from or merge into concepts, according as the mind is excited or calmed.

The intellectual sentiments are based on the concept of ourselves as student or thinker; the æsthetic, on the concept of ourselves as critics; the moral sentiment on the concept of ourselves as resolving or acting. This last sentiment may take as many forms as there are different ways in which we can act. The concept of ourselves as fulfilling the parts of citizen, member of a family, business man, friend, all pass into the answering sentiments when we are sufficiently moved. What form those sentiments should take is a question for the student of ethics, politics, or æsthetics, and not for the student of mind.

The religious sentiment gathers up into itself all the feelings we have been hitherto considering. This follows

THE FEELINGS.

of necessity. For it answers to ideas of the greatest abstractness and generality; and these are formed by the combination of simpler ideas. The feeling with which we regard our surroundings—the conditions of our life—as a whole is based on the several feelings with which we regard the parts of those surroundings.

112. **Catalogues of Emotion.**—No attempt is made here to exhaust the list of possible emotions for the reasons already given (§ 100, end). Should the reader desire to follow the matter further, he will find the descriptions given by Aristotle in the second book of his *Rhetoric*, and by Spinoza in the third book of his *Ethics*, very suggestive. The last part of Mr. Spencer's *Principles of Psychology* is very useful, although it implies a view of heredity which is meeting with much opposition (§ 18).

CHAPTER IX.

WILL.

113. Scope of this Chapter.—In psychology we include under the name will much more than is ordinarily understood by it. By will is ordinarily understood that power which we possess of representing to ourselves some state of thought, feeling, or activity as an end or aim, and of carrying out the movements of our bodies or the processes of thought by which we may attain it. But we are about to consider involuntary as well as voluntary action, and shall take account of acts like sneezing, in which there is no implication of purpose, as well as the more complex processes of will involved, for instance, in forming a resolve and carrying it out.

114. Physiology of Action.—The way in which the motor nerves are connected with the other parts of the nervous system has already been described (§§ 10, 13). But motor nerves are not only connected with other nerves—they are also connected with muscular tissues; and when they are excited they cause these muscular tissues to contract or shorten. All the movements of the body, elaborate and complicated as they may be, are caused in this simple fashion by the contraction of certain muscles.

A large part of our movements are caused by motor

impulses of which we are conscious. At this moment my hands are moving in obedience to such. By motor impulses we also check movements into which our limbs tend to fall, as when we try to restrain the second and fourth fingers from moving when the third finger is raised.

There are other movements, which may indeed follow upon motor impulses of which we are conscious, but they go on equally well in the absence of such conscious effort, and sometimes in spite of it. Sneezing, coughing, yawning, breathing, are all acts of this class.

Lastly, many movements are carried out without our ever being conscious of the motor impulses which give rise to them. Such, for example, are the movements connected with the processes of circulation and digestion.

115. **Reflex Action: Instinctive Action.**—Movements are said to be voluntary when they follow upon an effort —a motor impulse—of which we are conscious. We must carefully distinguish such cases from those in which we are merely conscious of the muscular sensations which follow. We are often conscious of blinking, but we very rarely put forth the motor impulses which cause it (§ 30).

Those actions which are thus produced without the intervention of a conscious motor impulse are sometimes marked off from voluntary acts under the name *reflex*.

The simplest reflex actions are those in which a single contraction follows upon a single sensory excitation. As an example we may take the closure of the iris which follows upon a bright light-stimulus. The regulation of many of the physiological processes is brought about by similar reflex actions; as the effect of one movement passes away, there arises a state of disturbance, which acts as a stimulus to the movement which will redress it. To abstain from breathing causes a state of uneasiness which gives rise to the next in- or ex-piration.

Most reflex tendencies with which the child is equipped at birth, or which unfold themselves soon after, are characterized by a kind of purposefulness; the start caused by the application of a hot object causes us to escape a burn; the closure of the iris protects the retina from undue stimulation by light. Actions like these, by which without conscious purpose on our part we are protected from harm or assisted to some desirable end, are sometimes marked off as *instinctive*. Such actions are displayed in a far more marked manner by many of the lower animals. Ichneumon flies, for example, lay their eggs inside caterpillars, and the larvæ feed on the fatty portions of their bodies, avoiding the vital parts. Now it is scarcely to be supposed that these larvæ understand the anatomy of the Lepidoptera; hence we must imagine that in selecting their food they act in a reflex manner, turning aside from certain parts in response to stimuli which their nervous system transforms into movements of the needed kinds. We may regard instinct, then, as a property of the nervous system by which certain stimuli are enabled to produce movements of a somewhat definite and complicated character. It is thus, in other words, complex reflex action.[1]

In proportion as a creature is equipped with instincts, that is, in proportion as external stimuli set up movements of which it is not the conscious originator, so does its organization possess the character of a machine. It is here that we find so great a difference between human actions and those of the lower animals. Speaking generally, all the more elaborate series of movements performed by the latter

---

[1] Spencer, *Psychology*, part iv., c. 5.

seem to be performed in the same mechanical, and therefore regular, way, as these natural processes to which we do not attach the idea of consciousness at all; the hexagonal chambers formed by the bee for the reception of honey are as regular as the hexagonal crystals sometimes found in quartz. With man, on the other hand, instinctive action is almost confined to the regulation of physiological processes. Hence the young child has to acquire by practice powers of action which other creatures possess from their birth. The chick begins to walk about and to peck at its food immediately it is out of its shell; the human infant is almost without command over its own limbs. Darwin says of one of his children—" The movements of his limbs and body were for a long time [after birth] vague and purposeless, and usually performed in a jerking manner."[1] But this helplessness which seems at the first blush to handicap the child so seriously in the struggle for life, is a presage of a capacity of movement which in its variety and range is excelled by no other creature. The child starts with a plasticity not only of mind but of muscle, which can be moulded into the most diverse habits.

116. **Random Action.**—Reflex action, whether in its simpler forms or in the more complex forms marked off as instinctive action, is executed without any clear idea of the result being present to the mind. We now come to a class of movements, namely, random movements, which stand half way between the unconscious and purposeless performance of reflex actions and the conscious and intended acts called voluntary.

Random movements are caused by a state of general nervous excitement, and we have a vague idea of the relief

[1] *Mind,* vol. ii., p. 286.

to be brought by them. They may be due to the application of more external stimulation than can be responded to at once in any definite manner, or they may express the disturbance of the nervous centres to a greater or less extent..

Strong external stimuli give rise to random movements, as when certain skin diseases, by irritating the sensory nerves from the skin to an unusual extent, throw the patient into convulsive movements.

A more characteristic kind of random movement is caused by the state into which the nervous system falls, after being comparatively unexercised for some time; the processes of nutrition overtake those of waste, and the nerves fall into an unusual state of readiness to act, such that the slightest impulse calls forth a relatively large amount of movement, whereby the pent-up energies are given an outlet. The aimless waving of the child's arms, as it lies in its cradle and its general restlessness, which does not decrease as it grows older, are the efforts of its accumulated nervous forces to relieve themselves. The fidgeting which comes over a large assembly, towards the end of a long address, is due to a similar state of things in the mature organism.

Lastly, most forms of emotional expression (§ 99) are caused in the same way; disturbances in the higher nervous centres give rise to muscular disturbances which are not directed to definite ends.

In random movement, then, the general stimulation of the sensory nerves, or the general excitation of the higher centres, gives rise to correspondingly general muscular contractions. We are usually conscious of the nervous excitations which precede such contractions; these often announce themselves to us as a feeling of uneasiness which may rise to positive agony through all the scale of pain. In a similar fashion strong feelings of pleasure may cause random movements.

117. **Motor Representations.**—At first, apparently, there is no direct connection between the idea of certain movements and their performance. We can still perceive this, even in later years, when we attempt to perform some actions under slightly changed conditions, as for instance in tying some new form of knot before a looking-glass; we are unable to put forth the requisite motor impulses. The question now is, how do we come to associate each particular motor impulse with the idea of the particular movements to which it will give rise? How do I know that a certain effort will move my hand to my face?

This involves in the first place that we have clear ideas of our different movements, and in the second place that each such idea is associated with the corresponding motor impulse.

The reader may easily observe for himself how sensations of touch and of sight combine with muscular sensations when we watch our own limbs in movement (§ 81). Such a combination forms the mental picture of the answering movement. As the child watches the changed postures its arms and legs take from time to time, it gradually gains a number of these complex ideas of movement. This process is accompanied by a process of association, by which these ideas are attached to particular motor impulses. It finds that by one kind of effort it can move its hand to its face, by another, its hand to its foot. And thus by degrees it becomes aware how each particular movement can be carried out by some motor impulse. In this process reflex movements help the child a great way. The start back, for instance, which it executes on striking its hand against some solid object, becomes so associated with the sensations of resistance that they recall it when they are felt again. The motor impulse, of which at first the child was not specially conscious, slowly comes to view. The attention is directed to this particular movement; and the child gradually disentangles the motor impulse from the associated ideas. Random movements similarly make the child's powers known to it; the vague impulses, which give rise to vague movements, gradually become clear and distinct; definite impulses become associated with definite movements.

By these means, there is slowly formed a picture of all the kinds of movement of which the child is capable, and side by side there comes the growing consciousness of the power to put forth the answering motor impulses. This idea of self, as the possible source of a great variety of movements, runs through all our conscious states. But the process of acquisition is a slow one, and is never carried to its complete issue. Mr. Darwin remarks of his infant son—"It was surprising how slowly he acquired the power of following with his eyes an object if swinging at all rapidly; for he could not do this well when seven and a half months old." Every one indeed learns the more ordinary uses to which his limbs may be put; when, however, the older child or the adult attempts to increase the range of his movements, he has to undergo the same kind of apprenticeship as the child in learning to walk or to grasp. The careful adjustment of effort to stroke required in playing billiards, the ordered alternation of movements required in swimming, the elaborate combinations of movements of wrists and fingers required in playing the piano, are alike attained by the association of particular impulses with the ideas of the requisite movements. And those adults who know how difficult these later acquisitions are, may be enabled to comprehend the amount of effort put forth by young children in learning the, to them, not less novel performances of handling and locomotion.

By repeated practice, however, the requisite motor impulse becomes so connected with the idea of each movement, that by the mere thought of the movement, the motor impulse is called up which produces it. This fact lies at the basis of imitation, and will be considered in the next paragraph.

118. **Imitation.**—In the absence of competing impulses, or in the presence of weak ones, the clear idea of any movement will sometimes give rise to the answering motor impulse, and so to the movement itself. Hence we find that children and adults of a low development, whose minds are but slightly equipped with ideas, and so are less exposed to the competition of impulses, idly fall into any activity which is suggested to them. Children's games consist very much in imitating the actions of their elders, partly of course as a means of flattering their sense of importance, but not altogether so. And it is the same with savages. The Fuegians "are excellent mimics; as often as we coughed or yawned, or made any odd motion, they immediately imitated us. Some of our party began to squint and look awry; but one of the young Fuegians . . . . . succeeded in making far more hideous grimaces. They could repeat with perfect correctness each word in any sentence we addressed them, and they remembered such words for some time. Yet we Europeans all know how difficult it is to distinguish apart the sounds in a foreign language."[1]

The clear motor representation on which action thus

---

[1] Darwin, *Journal of Researches*, chap. x.

follows is naturally most frequent when it is suggested by some immediate *sense impression*, such as the sight of a movement, or the hearing of words or other sounds. The term imitation is usually confined to such cases. But the process is radically the same when the mere *idea* of a movement calls up the answering motor impulse. Let us single out two typical cases.

The *word of command* shall be the first. When the drill-sergeant first tackles a raw recruit, he first shows him the actions and positions into which he must fall, and then attaches to each group of actions and to each position a sign, on hearing which the recruit is to execute the movements symbolized. When the recruit has succeeded in combining the idea of the required movements with the answering symbol, we have a case in which the mere idea of a movement may be directly followed by the necessary motor impulse.

Sympathy in its first beginnings offers another example of the way in which the idea of a movement passes into the effort which produces it. When the child becomes able to represent to itself the impressions and feelings produced in its fellows by the course of events around them, it proceeds to the thought of the actions by which itself would respond to such experiences; the thought of the pain undergone by another leads to the shudder and the cries which such pain would produce if actually felt (§ 108).

119. **Attention and Voluntary Movement.**—Just as impressions and ideas must be clear and distinct, in order that they may be grasped in their bearings one upon another, so the ideas of movement and the motor impulses attendant upon them must be clear and strong, in order to pass into action. Here we have a typical case of the activity of the

attention (§ 42). We shall understand, then, by a voluntary movement, a motor impulse preceded by a clear idea of the movements to which it will give rise.

When there is only one set of ideas of movement before the mind, and these are very clear and strong, we perform the answering movements; such for example are the cases considered in the last paragraph. If the mind is weak by constitution or by disease, and consequently can hold side by side but few ideas, any strong idea of movement may pass into the action itself—suggestion has a specially strong effect. The phenomena of hypnotism may probably be explained in this way: the mind of the subject is thrown into such a state that each suggested movement tends very strongly to realize itself. Ideas like these, which tend to pass into action without any necessary connection with some end to be gained by their means, are called *fixed ideas*. Such ideas, generated by seeing or having suggested the means to action, explain the unpleasant ways in which individuals overflowing with energy find vent for it; the overpowering desire to disturb the silence of some great meeting, to interrupt a singer, speaker, or actor, has probably occurred to most of us at some time or other. Practical joking is an example of more or less acute insanity of this kind.

In the healthy individual, phenomena like these are confined to those earlier years in which the mind is but scantily equipped with ideas; children are more prone to imitate and to play practical jokes than their elders.

Voluntary movement takes on a slightly more complex shape, as experience leaves more and more traces behind it. It is no longer a question of any single motor representation passing into reality, but of one among many competing representations. Through the conflict of impulses which thus arises, the mind is led to perform some of its most characteristic and important operations, such as choice, decision, resolution, and so forth.

120. **Trains of Movement: Habit.**—When we have performed a simultaneous combination or series of movements on several occasions, they leave a trace in us, such that it is more easy for us to perform the same movements again. Hence, in order to gain the power of performing a certain set of movements we practise them one after another. In this way, groups of movements are built up like groups of thoughts or feelings. By frequently performing movements together, simultaneous combinations are formed; thus in swimming the arms and legs have to be drawn up and stretched out simultaneously. Again by frequently performing movements one after another, serial combinations are formed; in waltzing, for example, a certain series of movements has to be practised until it becomes habitual. In cases like this, each movement calls up the next and so on to the end of the series. Pianoforte playing affords an excellent example, both of the combination of simultaneous movements as in striking the chords, and of successive movements in playing a piece from beginning to end. It is by this constant repetition of movements, that we obtain that command over our limbs, by which we are enabled to perform the various actions called for from us, or purposed by us, from moment to moment. Walking, running, jumping, swimming, riding, rowing, writing, playing on some musical instruments, drawing, and so forth, are all combinations of movements which we learn to carry out by repeated practice.

Let us now observe one or two characteristics of these habitual actions.

In the first place they are performed more *speedily*. The child learning to write forms but one stroke or letter at a time; at a later date he will write a word in less time than now it takes him to form a letter.

In the second place they are performed more *accurately*. A fair penman, provided that he does not write beyond a certain speed, will form each letter more neatly, and preserve the alignment of his words far more closely, than he could at first, even with the greatest care.

In the third place, habitual actions are performed with but little call upon the attention; neither the motor impulses which set them going, nor the muscular sensations which announce their execution, rise to view; they are performed *almost unconsciously*. The action of walking demands at first the child's undivided attention; in a few years constant practice will have enabled it to hold a conversation as it walks, without having to make an effort at each step, or being conscious of the muscular impressions which the act of walking produces in its legs and feet.

In other words, actions habitually performed become reflex in the sense referred to on p. 199; on the appropriate sensory impressions being received, they are executed without the co-operation of the attention, and sometimes in spite of it. Prof. Huxley quotes the anecdote "of a practical joker, who, seeing a discharged veteran carrying home his dinner, suddenly called out 'attention!' whereupon the man instantly brought his hands down and lost his mutton and potatoes in the gutter."[1] This action followed as directly

[1] *Physiology*, p. 302.

upon the sensory impression (the sound), as the start which follows upon touching a hot object.

Actions of this class differ from other reflex actions only in once having been voluntary and in being more complex. By habitual actions taking on themselves this reflex character, an economy of the mental resources is effected. Series of actions, each of which involved at first a separate act of will, become so associated together that we need only to will the first, and that calls up the second without any further special effort, and so on to the end.

The fact that actions, each of which depends on a voluntary effort, are performed less rapidly than habitual actions, is partly due to the fact that each special act of attention involves time (§ 42).

121. **External and Internal Suggestion.**—Ideas of movement are not only associated with one another; they are associated with impressions and ideas of other kinds. Hence they may be called up by these other impressions and ideas. The child finding that by lifting its hand to an apple tree it can get an apple, the sight of the apple becomes associated with the ideas of the movements necessary to reach it; the sight of the apple thus becomes a suggestion to that particular series of movements.

Suggestions to movement may be distinguished, according as they are due to impressions such as the visual impressions in the case just considered, or to ideas.

Suggestion by impressions may be marked off as external. This chiefly consists in the impressions which *external* objects affect us with; the stimuli we are constantly receiving set us moving in various ways. The sensations we receive from the muscles are also a source of this suggestion by means of impressions.

Movements are also suggested by ideas; this may be called *internal* suggestion. When we act in obedience to some idea which cannot be realized at once, we are acting in response to an internal suggestion.

Children are at first almost entirely under the influence of external suggestion; each new impression impels them to some new action. It is only as they grow older that the springs of action become internal to any great extent.

122. **Suggestion of Trains of Movements.**—Most of the actions we perform are parts of series of movements which have become conjoined by habit. Hence when we speak of movements as being suggested, what is usually meant is this—that some *series* of movements is suggested. Now it is rarely possible to hold in the mind the complete pictures of trains of mental operations, and so the first member in such trains is taken as representative of the others. Hence in speaking of the suggestion of trains of movements, it is not meant that each series of actions suggested is followed out in thought to its close, but that the tendencies to perform the first acts now of one series, now of another come to mind. Just as the first thought of a series becomes representative of the rest, so the first movement of a series becomes representative of the succeeding movements. Thus the attitude taken by a diver the moment before he makes his plunge suggests very strongly both to himself and to the onlookers the movements involved in the plunge.

123. **Entrance upon Fresh Series of Movements.**—From time to time we come to the close of some series of actions, and we are called upon to decide what shall follow them. The boy who has finished his lessons for the next morning, has to decide whether he will go a walk or stay at home and play at some game. As engagements accumulate in

later years, the completion of the series of acts entailed by one engagement only lays us under the obligation to go on to fulfil the next, and we are precluded from following out fresh impulses. Sometimes it almost seems as if we were machines going a defined round of work, without ever the opportunity of exercising a free choice. In these cases, it is not so much the conflict of impulse that is in question, as the obedience to the next duty.

These points at which we are called upon to start upon new series of actions form turning points, crises, of more or less importance. We are called upon to make some decision, and the temper in which we make it determines whether we shall be carried right through the chain of actions thus inaugurated, or whether we shall be turned aside to some new course of action before we have completed the old one.

124. **Motives.**—Let us consider the general state of mind which immediately precedes our entering upon some action or train of actions. We have already considered the part played by motor impulses in particular (§ 116).

The amount of action which is due to conscious motive is comparatively small; we have to put on one side reflex actions of all kinds, from the simplest to the most elaborate, including those which have become reflex by habit.

The simplest kind of motive is that which gives rise to imitative movements; the mere sight of a movement leads to its execution, is in fact its motive (§ 117). But as the mind gathers experience it can rarely be impelled to act by so simple means as this. So many ideas of movement are present to it more or less distinctly, that none of them can master it unaided. In other words the idea of a movement, in order to pass into action, must be supported by other ideas of more or less strength. And it is the mass of ideas thus constituted which forms the motive to the action.

Motives have power over us in proportion to the power of their constituent elements. Since pleasant and painful elements are among the strongest which enter into our states of mind, it has sometimes been thought that all motives must consist in the pursuit of pleasure or the avoidance of pain. But ideas may also become strong through their contrast with other ideas; the shock of surprise and the related emotions are among the strongest motives (§§ 31, 100).

Then again, if we are determined on several occasions by certain motives to act in certain ways, we may come to do so habitually. The casual association of ideas is not only responsible for habits of thought; it is also responsible for habits of action. We may fall into the habit of deferring to the judgment of certain persons or certain newspapers, and this too without any compulsion being exercised upon us. It becomes a sufficient motive to a given action to know that it is recommended by our habitual advisers.

Every time that we respond to a given motive makes the motive more powerful over us. Thus, motives which are not very strong in themselves, by being acted upon very often, ultimately become more effective than others of a greater intrinsic interest. Most persons know what it is to lose some great opportunity, or to accept some great sacrifice which might easily have been averted, through a disinclination to break through an easy routine or to lose a familiar gratification.

In proportion as motives become habitual, however, they cease to be motives; the occasion which suggests the action is scarcely observed, and we fall into the old habit without any activity of the attention. We act mechanically.

125. **How Motives Pass into Action.**—Motives may pass into action in two ways.

In the first place a group of ideas may come into the mind, and by the process of association may bring up an idea of some movement or series of movements which is forthwith realized. The sight of a stranger leads to the idea, and then to the act, of throwing half a brick at him.

In the second place some idea of movement may be present to the mind, but is not strong enough to pass into action until it is supported by a number of associated ideas. We may think of performing some action, and may hesitate until a number of considerations occur to us which would lead us in its direction.

126. **Conflict of Motives.**—Corresponding to these two cases are two ways in which motives may conflict with one another.

In the first place, the masses of ideas which ultimately lead to action may compete with one another for the supremacy, and it may not be a question at all of any idea giving rise to movement until the victorious idea brings it about. This is simply a conflict of ideas (§ 52).

In the second place, a number of motor impulses may tend to pass into action at the same time. We may be in a state of excitement, turning now this way, now that way. Now one idea of movement tends to pass into action, now another. When we are prevented from following out such a motive to action, it is usually because some mass of associations prevents us. The child reaching out its hand to take an apple is checked by the thought of the associated punishment.

127. Concepts as Motives and Ends.—The reader has already been asked to consider concepts as passing into emotions (§ 101); they also become motives. Few, if any, concepts are without motor elements, and when the concepts become very clear and strong, their motor elements also become very clear and strong and tend to pass into action.

Thus it is that we like to visit those places in which we have undergone any moving experience, provided of course the experience was not so painful as to inspire us with a positive aversion for its circumstances. If we are in the neighbourhood the idea of the spot becomes clear, and with it the idea of the movements by which we reached it before (and may reach it again), and these ideas becoming stronger pass into action. Here we have an instance in which the presence of an idea makes for its own realization. Ideas also sometimes impel us to the movements by which we escape realizing them. The idea of a place in which we have undergone some repulsive experience, brings up more clearly the movements by which we may escape the detested spot. But sometimes, if the idea of the obnoxious experience is very strong, it may lead us by a kind of fascination to realize it, although we are conscious that we are about to undergo something painful and abhorrent, as when a murderer haunts the place of his crime.

The cases just considered are those of clusters of ideas—concepts—attaching to particular objects. When we come to concepts answering to whole classes, like horse, book, &c., the actions they suggest are vague and general. But even here we can trace the motor suggestions; the concept horse suggests the movements of riding or driving, the concept book suggests turning over the leaves and reading. Those concepts like home, country, business, religion, which depend on ideas nearly always present to us, are motives from whose influence we are rarely free for long together.

Concepts then are capable of becoming motives in so far as they contain motor elements. Even cases which seem to contradict this, are probably to be explained in the same way. The child cries for the moon as it would not do if it knew that it was a quarter of a million miles away; it imagines that the moon can be handed down to it by its nurse, like the indiarubber ball on the mantelpiece. Here the idea of the moon as something to be handled is a motive.

Further, motives are strong in proportion as the idea of ourselves enters into them. We never desire anything in a quite abstract fashion; it is not pleasure that we seek or pain that we avoid, but we represent ourselves as being in a state of pleasure or pain, and these ideas suggest movements by which the state is realized or escaped.

Hence we do not desire external objects for their own sakes, but as entering into some pleasant or painful state of our own; that is to say, such objects are sought or avoided through being associated with our own pleasant or painful states of mind. It is only metaphorically, for example, that money can be called an *end;* the end is the state of possessing it.

128. **Dependence of Action upon Feeling.**—Action is specially dependent upon feeling. It is only as our ideas rise in intensity, that is, as they take upon themselves the character of feeling, that they are capable of becoming motives to action. Thoughts which, at their first presentation to us, stand in little or no connection with our life, have no influence upon us; they neither suggest nor prevent action. As, however, they become rich with associations clustering round them, their intensity increases, and consequently

their power of determining what we shall do. Thus every circumstance which increases the intensity of a thought renders it more able to move us to action. In other words our thoughts become motives by taking on an emotional character.

129. **Dependence of Action upon Thought.**—Action indeed demands that our thoughts shall take on an emotional character. Many purely intellectual states come before the mind without in the least moving it to action.

There is a special sense, however, in which action depends upon thinking as an intellectual process; we are rarely able to attain a desired state of mind, or to escape one for which we entertain aversion, without some definite effort of our own; we must perform some particular series of actions. Hence we must observe the circumstances amid which we are about to act, and adjust our movements to them.

The way this comes about is as follows—When an idea is very clearly present to the mind, its various constituent elements, and the relations between them, rise to view (§ 41). The feeling of hunger is usually associated with the ideas of the movements by which we can obtain food; it is connected, for example, with the idea of going home for dinner, with all the circumstances of time and place, and the movements of walking, &c., implied in that idea. Unless this takes place, unless the feeling which impels to action maintains somewhat of its intellectual character, it is not a motive. The feeling of hunger must be associated with the idea of the movements by which we may satisfy it, or else it does not impel us to any definite action. When an impulse is thus devoid of any idea of the means to its realization, it may indeed impel to action; but the movements will not be any manifestation of will; they will simply be the random expression of emotion. The actions of a person overwhelmed by some strong emotion of sorrow or pain show this purposeless striving.

It thus appears that a state of excessive emotion defeats its own ends; the too eager pursuit of one thing, or the too anxious avoidance of another, upsets that mental balance which is necessary to regular thinking. Thus, to be surrendered unreservedly to a narrow current of feeling, is to interrupt the processes of will no less than of thought. The unwavering pursuit of any single aim defeats itself; only a combination of aims can offer the mind the necessary scope for its activity.

130. **Means and Ends: Subordinate Motives.**—The child's actions at first are performed, we said, in obedience to the stimulus of the moment (§ 121). That is to say, the series of actions which it enters upon are continually being interrupted by some fresh suggestion. As it grows older, and its experiences increase, it becomes more and more capable of entering upon long trains of actions. Thus, it is a great step when it begins to look so far forward as the next day, and to prepare its lessons over-night; the series of actions thus begun are only completed when it takes up its lesson next morning. It is the same with other series of actions; they are only complete in themselves when we reach the last stage. The pursuit of an occupation by which we earn a livelihood, no less than the repetition of actions by which we acquire particular powers, consist in trains of actions which lead up to some last stage, some *end* which we desire to attain.

The importance of these series of movements depends, therefore, upon the fact that we rarely attain a desired state by a single movement. Even in so simple a matter as walking down the street, our object, to be at the other end, is only attained by a series of movements.

In more complex cases, the means to the desired end may consist of long series of movements and of intellectual operations combined in various ways; the operations by which a stockbroker tries to produce a rise in prices are such.

Observe that in this case, intellectual operations express themselves in movements—writing letters, giving orders, etc.; hence we are still considering a series of movements. It is only when the desired state is purely intellectual, and does not depend on external conditions, that muscular contractions do not enter into its production. To think out a difficulty may be a long business, and yet not involve the movement of a single muscle.

Observe also, that just as objects associated with our being in a desired state are sometimes called ends (§ 127), so objects associated with the production of that state are called means. Really, however, all such external objects are means, and not ends.

In all cases, where a series of intermediate steps intervenes between us and the realization of an end, it is the one nearest to our present state, and furthest from our desired state, with which we must begin. It is obviously necessary that this first step must be in our power, else we cannot go into the second and so on to the end. Hence when the mind occupies itself about some desired state, which cannot be immediately realized, the idea of that state can only become a motive to action when some train of ideas is hit upon which connects that desired state with the performance of some immediate action. It is to this first step that our immediate impulse is directed. When that is performed, we go on to the second. The motives that impel us to these interniediate actions, are the ideas of ourselves as being in the state produced by them, and these ideas receive the emotional colouring necessary in order to dominate us, by association with the main motive to the whole series. Thus, the main motive which started us on the series develops into the subordinate motives which lead us from one step to the next; the idea of ourselves as being in the desired state suggests the idea of being in the intervening states.

If these series consist of many repetitions of the same kind of act, as practised by Mr. Podsnap and his disciples in *Our Mutual Friend*—getting up at eight, and shaving close at a quarter-past, breakfasting at nine, going to the city at ten, coming home at half-past five, and dining at seven—then the motive need only be present to us on the first few occasions, in order to suggest the performance of the actions in question. Afterwards they become habitual, and we lose consciousness of the motive. It is only in series of actions, each one of which is different, and is not necessarily suggested by the preceding, that consciousness of the motive, or end in view, need be maintained. Thus, routine effects a great economy of our mental resources by making certain actions habitual. It follows that in order to perform series of actions of greater complexity than usual—series, every step in which demands a different treatment of our circumstances—demands are made which can only be satisfied by minds of more than average calibre, and overtask those minds which are equal, but not more than equal, to routine work.

131. **What is meant by a Volition.**—Summing up the results to which we have now attained, we may define

an act of will or *volition* to consist of the following elements—

(*a*) The idea of ourselves as being in some desired state.

(*b*) The rising to view by a process of association of the means by which that state can be attained.

(*c*) Our passing into some answering activity.

132. **Desire and Aversion.**—We may be scarcely conscious of the motives which impel us to an action at the time of performing it; this is the case with reflex actions of all kinds, including those which are become reflex by habit. While our actions are done consciously—that is, not in a reflex manner, or under the influence of habit—the motives to perform them remain clear and distinct. The emotional colouring of the motives also remains clear; and we can observe the elements of pleasure, or pain, which constitute the largest proportion of that emotional colouring.

The terms desire and aversion are applied to states of mind of all degrees of emotional colouring, from those of the most intense emotion down to states in which we act almost mechanically, and without consciousness of motive.

The term desire is applied to those motives in which the emotional colouring chiefly consists of pleasant elements.

The term aversion is applied to those motives in which the emotional colouring chiefly consists of painful elements.

The association of pleasure with any ideas increases their force as motives very greatly, and this, too, without necessarily disturbing the balance of our intellectual processes. The thought that some great enjoyment is offered to us in a certain direction, marvellously quickens our perceptions and reasonings.

Just as painful sensations are somewhat confused in character (§ 24), so the association of ideas of pain with other ideas tends to interrupt the current of thought. Hence the actions which follow upon a stage of aversion to anything, are not always directed to escaping it. It is only when we maintain a certain measure of self-control, that we can set about escaping from the state to which we entertain aversion. It sometimes happens that a state of strong aversion paralyzes us, or even drives us to the very state we dread so much (§ 127).

In desire, the idea of the state which we wish to attain, as it becomes more intense, gradually brings up with it the ideas of the means by which we may attain it; in aversion, ideas are brought up both of the way in which the experience may be avoided, and in which it may be undergone; both these are matters of interest, the one as offering an escape from the object of our aversion, the other as leading to it. And the latter, by its very strength, may act as a motive to action, although repugnant to us.

133. **Intellectual States as Objects of Desire.**—When several trains of thought compete for the domination of the mind, associations with pleasure or pain often determine the issue; some state of thought, some train of ideas, appears more attractive than the rest. In proportion, as this power of any train of ideas takes on more and more of an emotional colouring, so does it present itself to us as an object of desire or aversion. As it becomes stronger, so are we impelled to its realization.

This realization may not involve a single movement; we can ponder some perplexity and desire to solve it, and after some reflection may succeed in doing so. In proportion as our aims are directed to these ideal ends—in proportion, that is, as we seek to realize ourselves not in any external interest (the possession and use of tangible objects), but in some order and balance of our mind in and for itself, so do we enter upon that inner life which is the highest human interest, because it is the unique characteristic of human life.

134. **Deliberation, Hesitation, and Resolution.**—Where the impulses present to the mind at any moment harmonize, that is, when there is associated with each of them the same movement of the body, or the same direction of the thoughts, we fall into the activity suggested after the interval needful for the desire to rise into the central region of consciousness (§ 42).

The fact that our impulses are in conflict is usually brought home to us by the feeling that, first one action and then another, which we are upon the point of performing, arouse controlling impulses which prevent them from being carried out. Side by side with each impulse, the ideas more or less clear of the movements associated with them, or of the thoughts which would follow, come to mind. This rising to view of conflicting impulses, and the accompanying motives, constitutes the process of *deliberation*.

If this process is often repeated without any impulse or motive gaining the supremacy, the state of mind is one of *hesitation*. Each impulse and motive, as it passes under review, overcomes the previous one, but its very rising to dominance tends to bring up another impulse or motive, which will exclude it in turn.

When at length the conflict terminates in favour of some single motor impulse or train of thoughts, we are said to *resolve*. It is only when the action about which we are deliberating is an immediate one, that the act of resolution necessarily makes itself manifest in some motor impulse. When the action is a future one, the act of resolution merely consists in the temper of mind into which we fall;

this is called a *purpose*, or *intention*. To intend, then, is to resolve with reference to some action in the future. This of course does not exclude some other motive from overcoming that which has obtained the mastery on the immediate occasion. One of the commonest mistakes made in our estimate of our own mental processes is to regard the passing influence which some motive has gained over us, say, in response to a stirring sermon or address, or to some moving event, such as a great sorrow, as its permanent value. It indeed overlays for the time our other motives, but as it fades away these reassert their power, and old tendencies come back all the stronger for their temporary suppression. To be unable to form purposes which shall maintain themselves over long periods of time in the teeth of contrary impulses, is the evidence of a mind in which ideas have not clustered together into those great and powerful groups which constitute the permanent principles of action.

A natural vanity makes us exaggerate the power of the purposes which we form from time to time. But this weakness is also our strength. Those persons who under-estimate the power of their resolves, already have the conditions at hand which shall further weaken their will. "When one who often anxiously wavered between hope and fear, was one day consumed with sadness, he prostrated himself in prayer in the church before a certain altar, and revolved these things within himself, saying, 'Oh, if I did but know that I should persevere on and on!' All at once he heard within himself the divine answer: 'And what wouldest thou do if thou knewest this? Do now what thou wouldest then do, and thou wilt be safe enough.'"[1]

135. Personality.—It may have seemed to the reader as

[1] *The Imitation of Christ*, Book i. chap. 25.

we have gone on tracing the origin of our ideas to impressions produced in us by external objects, and our actions in great part to reflex nervous processes, both innate and acquired by habit, that but little scope was left for the manifestation of those attributes which he regards as distinctive of his own being—personality and self-control. It remains, therefore, in conclusion to point out that, although the connection of cause and effect is as rigid and uniform within the limits of our own proper nature as beyond them, this fact rightly interpreted does not conflict with any of those attributes of our own individual life which we regard as, at once, the most certain and the most characteristic.

When we speak of laws of mind, what should be understood by the phrase is simply this, that the mind has certain fixed manners of operation; if it were not so, instead of having a nature, a constitution, of its own, it would be a chaos. There is no suggestion of compulsion from without in all this. The mind in acting according to mental laws unfolds its *own* nature. And the powers which it so manifests are only possible as they form part of a whole ; and this whole is *our* mind, our personality (§ 39).

Those very circumstances which, at first sight, seem to hamper the mind's free action really help it to a fuller dominion over its conditions. Let us see how this comes about :—

In proportion as our states of consciousness are clear and distinct, so do they exhibit most clearly the action of mental laws. But only very few elements can be present to us in a clear and distinct manner at any given moment. Hence the distinctive operation of mental laws is confined to a comparatively limited portion of our conscious life. So far as mental operations are reflex and automatic in character—

and by constant repetition our acts and even our thoughts and feelings take on themselves this character—so are they removed from the clearest current of consciousness. But this very automatism makes it possible for the few elements of which we are clearly conscious at any time to stand for a large number of others of which we are not thus conscious. Each element which rises to the clearest region of consciousness becomes associated with the traces of many past experiences; and so each of the few elements of which we are clearly and distinctly conscious stands for large series and clusters of such traces. Thus it is that our single ideas stand for many objects in the act of conception, and single inferences for many cases in the act of generalization (§ 67). Hence too, when we resolve to perform a series of movements, their execution has usually been made easier by many previous actions of the same kind, so that we need only think of the motive which brings up the idea of the first action to be performed, and the rest follow automatically (§ 122). "It is a general rule that in our voluntary actions we have before us the aim alone, and leave its execution in detail to a mechanism that is born with us or acquired."[1]

Thus over against the mind's proper activity, there stands the working of a mechanism which it forms by its own ideas, feelings and movements, which, from being voluntary, become automatic. On the whole the mind's past and present activities are so far harmonious that the free action of the moment is not hampered by the traces of preceding activities. Still there sometimes arises a conflict between present motive and habitual tendency—a conflict which is one of the salient features of the mental life. "The habituated flesh

[1] *Wundt*, Vol. ii. p. 500.

becomes the suggester of crime to the will which first constrained it to sin, and now wearily rebels against the habits of its instrument."[1]

136. **Conscience and Remorse.**—Our self consists in the state of consciousness in which we are at any time, combined with the capacity of passing into a succeeding state of consciousness. That is to say, our self consists at each instant of a group of thoughts, feelings, and impulses (§§ 32, 33). Now by the operation of the mind, impressions, ideas, feelings, and impulses are built up into great clusters and series, to which the mental elements of which we are immediately conscious are referred; and the consciousness of these great clusters and series rises dim in the background of consciousness. Thus we may regard such a group of ideas as being partly realized in our self of the moment (§ 111).

In proportion as such a group of ideas answers to the whole of our previous experience, so will it tend to occupy consciousness undisturbed by conflicting ideas; while a group of ideas which has not this warrant in the past and present will always be exposed to the opposition of groups of more authority. Observe further, that when a group of ideas is denied for any length of time its proper expression, when it is repressed, its struggles to re-enter and dominate consciousness become increasingly violent for a time.

Here we have the explanation of conscience and remorse. A man is conscientious in proportion as the group of ideas in which his motives mainly consist, and take rise, answer to the widest view of his own experience and those of his fellows; while a man who displays little conscience, is one whose dominant group of ideas is circumscribed within very narrow limits. When we hesitate between several courses

[1] G. A. Smith, *Isaiah*, i. 423.

of action, our ultimate decision is moral or immoral in proportion as the dominant motive depends, or does not depend, upon ideas with the deepest root in experience. The insight, however, which would enable us to choose aright, is always in danger of being perturbed by the more pressing solicitations of the moment; the immediate impression, often depending on conditions chiefly physiological, overlays and conceals the traces of past experiences; and unless the mind is schooled to habits of calm deliberation, the strong desire or aversion, inertia or restlessness, may cause action or inaction, which is immediately after felt to be a blunder, or worse. But the immediate impression, which was the main motive to the act, passes away very rapidly (§ 34); old ideals come back with a light all the more lurid for their temporary eclipse, and we are afflicted with a consciousness of the discord between our immediate action and them. The strength of remorse, like all reactions, is proportioned to the strength of the motives we have broken through. In other words, remorse is a state in which the mind tries to recover its equilibrium, in order to become its true self again, by a violent effort.

There is an important corollary to this principle, which is not always borne in mind. In obeying the promptings of the largest ideals, it is dangerous to refuse satisfaction to less important motives; to try to crush any side of the mind's nature is to maim the mind, or drive it to rebellion.

The larger Ends to which our life is directed as a whole—the realization of our self in various ways—are attained by various means, and by association these means take on the character of subordinate ends. In connection with them, a species of secondary conscience arises. Thus the persistence in business, or in housewifely care, depends on the presentation to the mind of the connected minor ends, means of livelihood and private comfort. And so the horror of the unbusiness-like, or of domestic mismanagement, is capable of attaining an almost moral character; witness Caleb Garth's state of mind when Fred Vincy showed him a specimen of his handwriting,[1] or Mrs. Poyser's virtuous contempt of Mrs. Chowne's housekeeping.[2] A common mistake is to treat obedience to a subordinate end as sufficient in itself to form a character. Temperance in drinking is highly desirable, no doubt; but taken alone, it forms a very meagre equipment.

137. **Self-control.**—Many of the conditions by which we are determined to act at any moment are removed from the control of the attention. But although it is sometimes very difficult for the mind to override these conditions, yet we can bring new conditions into operation, and thus strikingly modify our tendencies to action in the future. By lessening the number of the occasions on which we tend to follow out a train of action, by exercising ourselves in impulses of contrary kind and so strengthening them, by filling our minds with other thoughts (which, if not actually opposed to the old tendencies, are, by preventing them from coming to view, almost as efficient as if they necessarily excluded them), the heaviest chains of habit may be slackened, and at last quite unwound (§ 38). De Quincey's *Confessions of an Opium Eater* afford an interesting analysis of such a process.

Since the emotional colouring taken by an idea largely determines its power in prompting us to act, the government of the feelings is especially important here. By controlling

---

[1] *Middlemarch*, chap. lvi.
[2] *Adam Bede*, chap. xviii.

their expression, and by guarding against those states of body (exhaustion, etc.), in which we are unusually subject to emotion, we may strengthen ourselves against impulses which we desire to overcome.

Lastly, since motives do not come upon us unsuggested, we can control them by regulating the influences—surroundings, company, books, etc.—which would suggest them.

138. **The Practical Syllogism.**—There is a striking likeness between the logical syllogism (§ 66) and the process of Volition (§ 131). The will, so to speak, finds its general principle or major premise (*a*) in some desired end drawn from an universe of ends (§ 136); *e.g. I ought to study law-books.* The minor premise (*b*) states the opportunity (§ 129) by which (*a*) may be realized. *Yonder are the books, and I have leisure.* The conclusion (*c*) is an act, or an abstention, according as (*a*) commands or forbids. Here it is an act: *I begin to read.*

Most deliberate actions can be analyzed in this way. We have here a leading instance of the underlying unity of mental processes. Just as the universe of discourse furnishes us with the principles of thought, so the universe of our desires furnishes us with the principles of conduct upon which we act. Now it has been pointed out (§ 127) that concepts pass into motives. Hence the source of our thoughts, and that of conduct, are in the last resort one and the same.

INDEX.

The numbers refer to the pages; the italic numbers refer to the pages where definitions or explanations are to be found.

ACTION, physiology of, 198; reflex, instinctive, *199;* random, 201; and feeling, 216; and thought, 217
Aesthetic, *7*
Alcohol, effects of, 33
Analogy, 145
Anger, 19, 191
Aristotle, 6, 123, 191, 197
Association, 85; laws of, 107
Attention, characteristics of, 93; theory of, 95; measurement of, 95; time taken in, 99; active, receptive, 101; and novelty, 102
Auditory combinations, 159, 167
Autobiography, 10
Average type, 7, 20; faculty, 12, 44
Aversion, *221*

BAIN, on classification of feelings, 185
Bashkirtseff, Marie, 10
Belief, *138*
Binocular vision, 68

Body, states of, *15*
Brain, double, 33

CHILDREN, 11; soon tire of new toys, 103; laborious acquisition of muscular powers, 204; games of, 205; act on impulse, 211, 218; narrow sympathy of, 192; cannot conceive alternatives, 139
Class, *129*
Colour, 70
Colour-blindness, 71
Command, word of, 206
Common sensations, 59; combinations of, 154
Concepts, *126;* crude, *112;* classification of, 132; as moods, 189; as motives and ends, 214; stages in forming, 132
Conscience, 3, 227; secondary, 228
Consciousness, limited in scope, 88; clear region of, 95
Crime, 7, 20, 182

INDEX.

Crises, in development, 42; in action, 212

DAMARAS, 95, 142, 172
Darwin, 45, 117; method of, 144, 146; *Biography of an Infant*, 11, 201, 204
Deduction, *139*
Definitions, functions of, 134
Deliberation, *223*
Descartes, 3
Desire, *221;* intellectual states as objects of, 222
Development, and growth, 38; in structure, 38; partial, 39; factors in, 42
Doubt, *138*
Dreams, 7, 92

ELEMENTS, *5;* combined in states of mind, 79, 81; series of, divergent, intersecting, and convergent, 115
Emotion, nature of, 179; expression of, 183; physiology of, 181; classification of, 184; trains of, 188; suggestion of, 189; conflict of, 189; harmony of, 190; and thought, 191; selfish, 192; interpretation of, 194
English character, 49
Environment, 37; physical, 48; social, 50
Ethic, 7
Excitations, *22;* sensory, motor, *ibid.*

Expectation, 99, 120
Experiment, 11; relation of sensation to stimulus, 18; on touch, 66; on suggestion, 109; on perception, 175; on emotion, 182
Eye, description of, 68

FACTORS, *5;* in mental development, 42
Faculty, nomenclature of, 3; limit to, 41
Ferrier, 11
Fidgeting, 202
Filmer, 146
Fixed ideas, *207*
Forgetfulness, 118
Fuegians, 43, 139, 205
Functions, nervous, how ascertained, 28; of spinal nerves, 25; of medulla oblongata, 32; of cerebellum, 33; of cerebrum, *ibid.*

GALTON, 45, 95
Garth, Caleb, 229
Generalization, *132*, 144
Genus, general, *129*
Gesture, 9
Greek character, 49, 189
Gushing, 193

HABITS, 82; of memory, 113, 119; of reasoning, 146; of movement, 208; effects of habit, 209
Hearing, 75

INDEX.

Hegel, 93
Hegner, 44
Herbart, 3, 14
Heredity, 45
Hesitation, *223*
Hindu character, 54
Hoffmann, 44
Holmes, Wendell, 41, 47, 114
Huxley, 24 ff, 209
Hypnotism, 7, 11, 18, 110, 207

IDEAS, opposed to impressions, 80, *148;* with common elements blend, 116
Idiocy, 44, 106, 184
Iliad, 170
Images, positive and negative, 73; representative, *112*, 123, 144; typical, 124
Imagination, 118; constructive, 119
Imbecility, 44
Imitation, 205
Impression, *148;* of relation, 78, 82
Impressionists, 10
Induction, 13, *143*, 145
Inference, *139*
Insanity, 7, 19, 191
Instinct, *200*
Interest, *97*, 179
Introspection, *8*, 81, 152, 177

JUDGMENTS, nature and occasion of, 135; intensive and extensive, 136 ff

KANT, 3

LANGUAGE, 9; and concepts, 126; as emotional expression, 136, 194
Locke, 3, 105, 116, 123
Logic, *7;* and psychology, 128, 138, 175
Lombroso, 182

MAUDSLEY, 33
Memory, nature of, 83, 104; kinds of, 113; verbal, 113; diseases of, 83, 106, 113; as survival of the fittest, 115; systems of memories, 124
Mendelssohn, 14
Mental laws, *15*, 18; processes and products, 84
Method, psychological, of Difference, 15; of Agreement, 19; of Variations, 19, 20
Mill, 15, 38, 168
Mind, states of, *14*, 79; and matter, 8
Moll, 18
Motives, 212; relation of, to action, 214; conflict of, 214; subordinate, 218
Motor impulses, 77; representations, 203
Movements, trains of, 208, 211
Muscular sense, 62; combinations of, 154, 161, 162, 165
Musical, pitch, timbre, 75; concord, 76

INDEX.

NATURE, as factor in development, 42 ff
Nerves, motor, sensory, 22, 24
Nervous, system, 24; excitations, *22*
Noises, 76

OBSERVATION, external, *9;* scientific, 175
Operation, typical mental, 193

PAIN, kinds of, 61; expected, more intense, 100
Pathology, mental, 7
Perception, *149;* and sensation, 148; and attention, 149; external, internal, 150; of surroundings, 168; of distance, direction, 170; of size, number, 171; of solidity, force, change, 172; of succession, 173; of space and time, 174
Persistence, law of, 82
Personal error, *99*
Personality, 78, 91, 224
Phrenology, 13
Plasticity, *43*, 201
Pleasure, 60; and pain not the sole conditions of interest, 97; nor the sole motives, 213
Podsnap, Mr., 220
Poyser, Mrs., 229
Presentation, *84*, *149*
Preyer, 11
Psychology, *1;* relation of, to other sciences, 6, 17; descriptive, 13; classification in, 14, 81
Purpose, *223*

QUINCEY, DE, 229

REASONING, beginnings of, 123; trains of, 137
Recollection, *105*
Reflection, 177
Reflex, nervous action, 30; name given to movement performed without conscious motive, 199
Religious conformity, 54; sentiment, 196
Remorse, *228*
Representative, *112*, 185, 187
Reproduced states, nature of, 85, 88, 110
Resolution, *223*
Richter, 45
Robertson, Croom, 57
Rousseau, 113, 136, 176
Ruskin, 150

SELF, consciousness of, 152; ideas of, 192, 196, 227
Self-control, 91, 229
Sensation, nervous conditions of, 22, 24; quality, intensity, local character of, 56; interest, discrimination of, 57; height, threshold of, 58
Sentiments, 195, 196
Sharp, Miss, 78
Sight, 68

INDEX.

Smell, 67; combinations of, 161
Species, special, *129*
Spencer, 45, 65, 185, 197
Spinoza, 197
Standards, psychological, 7, 12
Stevenson, R. L., 192
Stimulus, *22;* ideas as, 121
Suggestion, 106; external, internal, 109, 210
Suicide, 20
Syllogism, *140*, 143
Sympathy, 15, 193, 195

TACT, 194
Taine, 11
Taste, 66; combinations of, 161
Temperament, 83, 191
Temperature, sense of, 64
Tendencies, mental, strength of, 88; balance of, 92, 222
Terms, use of, 2
Themistocles, 118

Touch, 63; combinations of, 156, 161, 162, 165

USE and disuse, effects of, 41

VISUAL sensations, combinations of, 157, 162, 165
Vivisection, 11, 28
Volition, *221*
Voluntary movement and attention, 206; time taken in, 98

WALLACE, 40
Waltz music, 160
Wardle, Mrs., 102
Weber's Law, 18, 58
Will, 78, 199; strength of, 224
Wundt, *passim*

YOUNG-HELMHOLZ theory of colour sensations, 71

THE END.